Rethinking Science

Questioning Evolution

and the Theories

Your Guide to the Current Critical Scientific Debate

By

René Lauritsen

I am sure you will find this book interesting and informative.

If so, please consider giving it a review on Amazon and perhaps mention it or link to it on your social media

Thanks, it will be highly appreciated

Find other books at

www.VirtusPublish.com

TABLE OF CONTENTS

Overview of the Book's Aim and Scope

Welcome to "Rethinking Science: Questioning Evolution and the Theories." This book aims to critically examine some of the most foundational and widely-accepted scientific theories, specifically focusing on evolutionary theory and cosmological theories about the universe's origins and structure.

The objective is to provide a comprehensive yet skeptical look at them. We will explore the contradictions, limitations, and problems that exist within these theories, many of which are often overlooked or glossed over in mainstream discussions.

THE RATHER IMPORTANT
MISSION OF THIS BOOK

T he mission of this book is not merely to present a series of facts or to regurgitate established theories. Rather, it aims to serve as a comprehensive guide that invites readers into the heart of a critical scientific debate. This book is conceived as a journey—a journey that will traverse the complex landscape of evolutionary theory, and the various methodologies that scientists employ to decipher the mysteries of life.

Our objective is twofold. First, we aim to provide a thorough understanding of the prevailing theories in these domains, offering a detailed exposition of their foundational principles, historical development, and current applications. I believe that a nuanced understanding of these theories is essential for anyone who wishes to engage in a meaningful dialogue about them.

Second, and perhaps more importantly, this book aims to scrutinize these theories through a lens of healthy skepticism. We will explore the criticisms, limitations, and controversies that surround these theories, drawing upon a wealth of scholarly research and expert opinions. Our goal is to highlight the questions that remain unanswered, the assumptions that are often overlooked, and the ideas that challenge mainstream thought.

In doing so, we aspire to create an intellectual space where questioning is encouraged, where differing viewpoints are not just tolerated but actively sought, and where the pursuit of truth is considered the highest aim. I invite you to join this endeavor, to suspend their preconceptions, and to engage with the material in a spirit of open inquiry.

Through this rigorous examination, we hope to contribute to a broader scientific discourse—one that values skepticism as a virtue and sees debate as the cornerstone of progress. It is our conviction that by fostering a climate of critical examination, we can collectively move closer to a more nuanced and complete understanding of the world in which we live.

THE INDISPENSABLE ROLE OF RIGOROUS CRITICAL EXAMINATION IN SCIENTIFIC INQUIRY

In the ever-evolving landscape of scientific discovery and understanding, the role of critical examination stands as an indispensable pillar. It is crucial to recognize that science is not a monolith of unchanging truths but rather a dynamic field that thrives on scrutiny, questioning, and, at times, radical re-evaluation. Theories and hypotheses, no matter how universally accepted or empirically supported they may appear, are not immune to the invaluable process of critical examination.

The act of questioning established theories serves multiple essential functions. Firstly, it ensures that existing theories are robust enough to withstand scrutiny, thereby solidifying their place in scientific understanding. Secondly, it exposes the limitations, inconsistencies, or even inaccuracies that may exist within these theories, providing a fertile ground for further research and refinement. Thirdly, and perhaps most importantly, it opens the door for revolutionary ideas that could either modify or replace existing theories, propelling the field into new directions and possibilities.

In this light, critical examination is not an antagonistic force against scientific progress but rather its catalyst. By challenging the status quo, we are not undermining science; we are enriching it. We are not closing doors; we are opening new ones. We are not sowing discord; we are laying the groundwork for a more nuanced and comprehensive understanding of the intricate complexities of life.

Therefore, as we delve into the various topics covered in this book, we will not shy away from asking difficult questions or presenting viewpoints that challenge mainstream thought. It is through this rigorous process of questioning and debate that science moves forward, and it is in this spirit that this book has been written.

Invitation to the Reader

I hereby invite you, the reader, on this intellectual journey. Whether you're a scientist, a student, or simply someone interested in science, I encourage you to approach this book with an open mind. Let's explore the limitations of what we 'know,' question the accepted narratives, and in doing so, expand our understanding of the complexity of life and the universe.

CHAPTER 1

THE BASICS OF

EVOLUTIONARY THEORY

Introduction

Welcome to the first chapter of this intellectually stimulating journey—a journey that aims to dissect, scrutinize, and critically evaluate one of the most influential scientific theories of our time: the theory of evolution. This chapter serves as the gateway to a comprehensive exploration of evolutionary theory, a subject that has not only revolutionized our understanding of biological diversity but has also permeated various other disciplines, from psychology to medicine and even to the realm of social sciences.

Objective of This Chapter

The primary objective of this chapter is twofold. First, we aim to provide an in-depth introduction to the foundational principles that underlie evolutionary theory. We will delve into the historical context that gave rise to this theory, explore the key mechanisms that are postulated to drive evolutionary change, and examine how the theory has evolved over time to incorporate new scientific discoveries, particularly in the field of genetics.

Second, and perhaps most critically, this chapter aims to present a skeptical viewpoint on evolutionary theory. We will not shy away from asking hard questions or from presenting criticisms that have been levied against various aspects of the theory. Our goal is to provide a balanced view that not only celebrates the achievements and insights that evolutionary theory has provided but also acknowledges its limitations, controversies, and the questions that remain unanswered.

By the end of this chapter, you, the reader, are equipped with a nuanced understanding of what evolutionary theory is, what it aims to explain, and why it has been both celebrated and criticized. We will draw upon a rich tapestry of scientific research, expert opinions, and academic discourse to provide a well-rounded view of the subject.

So, without further ado, let us embark on this intellectual expedition to explore the intricacies, marvels, and debates that surround the theory of evolution.

SECTION 1

WHAT IS EVOLUTIONARY THEORY?

Explanation of what evolutionary theory aims to describe

Evolutionary theory serves as a comprehensive framework that seeks to elucidate the intricate processes responsible for the diversity and complexity of life forms on Earth. At its core, the theory posits that all species of organisms have descended from common ancestors through a gradual process of change over an extended period of time. This transformation is not random but is guided by a set of mechanisms, most notably natural selection, which acts upon the genetic variation within populations.

The theory aims to offer a cohesive explanation for a wide array of biological phenomena, ranging from the similarities and differences in genetic material across species to the intricate ecological interactions that shape the natural world. It provides the underpinning for our understanding of how simple, single-celled organisms could give rise to the vast complexity of life we observe today, including the myriad forms of plants, animals, fungi, and microorganisms that inhabit virtually every conceivable environment on Earth.

Moreover, evolutionary theory extends its reach to explain not just the physical characteristics of organisms—such as their size, shape, and coloration—but also their behaviors, life cycles, and reproductive strategies. It serves as a cornerstone in the biological sciences, influencing other disciplines such as medicine, psychology, and conservation biology, by offering critical insights into the origins, development, and functioning of living systems.

Evolutionary theory aims to provide a unified, scientific explanation for the origin and diversification of life, grounded in empirical evidence and subject to testing and validation. It is a theory that has not only stood the test of time but has also been refined and expanded upon as new data and technologies have become available.

The scope of evolutionary theory in explaining the diversity of life

Evolutionary theory is often presented as a comprehensive framework for understanding the origins and diversification of species, serving as a cornerstone for a multitude of scientific disciplines that explore the complexity and diversity of life on Earth. However, it's important to critically examine the expansive scope that evolutionary theory claims to cover and the various facets of biology and life sciences it purports to inform.

Taxonomy and Phylogenetics

Taxonomy is the science/study of classification. Phylogeny is the science/study of evolutionary relationships between organisms. While evolutionary theory provides a structure for classifying organisms based on supposed evolutionary relationships, this approach has been critiqued for its reliance on assumptions about common ancestry and the interpretation of morphological and genetic data.

Molecular Evolution

Evolutionary theory attempts to extend its principles to the molecular level, aiming to offer insights into the genetic basis of inherited traits. However, the mechanisms by which genes supposedly evolve over time and contribute to the adaptability and survival of species are still subjects of ongoing debate.

Ecology and Ecosystems

Although the principles of evolutionary theory are often applied to ecological studies, questions remain about how accurately these principles can explain interactions between organisms and their environments, as well as relationships between different species.

Adaptation and Fitness

Evolutionary theory endeavors to provide a framework for understanding "fitness," or the ability of an organism to survive and reproduce. However, the concept of natural selection as the driving force behind the survival and proliferation of advantageous traits is still a matter of scientific scrutiny.

Human Evolution and Behavior

While evolutionary theory has been extended to explain human origins, behavior, and even culture, these explanations often rest on speculative grounds. The development of human-specific traits like complex cognition are still subjects of intense debate within the scientific community.

Medical Applications

In medicine, evolutionary theory is used to understand diseases and drug resistance. However, its effectiveness in predicting future public health challenges or the evolution of pathogens is still under question.

Ethology and Animal Behavior

Evolutionary theory attempts to explain animal behavior through natural selection. However, the origins of specific behaviors and their contribution to survival and reproductive success are far from settled.

Macroevolution and Microevolution

Evolutionary theory claims to encompass both large-scale evolutionary changes and small-scale changes within populations. However, the unified framework it proposes for studying evolution at different scales is still a subject of academic scrutiny.

Sexual Selection

Last but not least, evolutionary theory tries to offer insights into the puzzling and often elaborate behaviors and physical traits related to mating and reproduction through the concept of sexual selection. However, the validity of sexual selection as a universal explanation for such traits is still a matter of ongoing research and debate.

In summary, while evolutionary theory claims a vast and all-encompassing scope, each of its applications comes with its own set of questions and criticisms. It's crucial to approach these claims with a skeptical lens, as many are still subjects of ongoing scientific debate and scrutiny.

Skeptical View on What is Evolutionary Theory

While evolutionary theory has been widely accepted as a cornerstone in biological sciences, it is not without its critics and skeptics. One of the primary issues that skeptics raise is the broadness and vagueness of the term "evolution" itself. The word is often used to describe a range of phenomena, from small changes within a population over a few generations to the origin of new species and even the origins of life itself. This lack of specificity can make the theory seem like a "catch-all" explanation, which raises questions about its scientific rigor.

Moreover, the term "evolution" is sometimes applied in contexts that extend beyond biological systems, such as cultural evolution, social evolution, and even cosmic evolution. While these applications may be intriguing, they can dilute the scientific meaning of evolutionary theory and contribute to its perception as a universal explanation for a wide array of unrelated phenomena. Critics argue that this broad application can undermine the theory's credibility and precision.

Another point of skepticism comes from the scientific community itself. Notable biologists and philosophers of science have questioned the explanatory power of evolutionary theory. For instance, the late evolutionary biologist Stephen Jay Gould raised concerns about the limitations of Darwinian explanations for phenomena like "punctuated equilibria," where species appear to remain stable for long periods before undergoing rapid changes. Gould's critique suggests that not all evolutionary changes can be neatly explained by natural selection and gradualism.

Furthermore, the concept of "random mutation" as a driver for evolutionary change has been scrutinized. Critics argue that the likelihood of beneficial mutations leading to complex life forms is astronomically low, making it an insufficient explanation for the diversity and complexity we observe in biological life.

In addition, the Modern Synthesis, which attempts to marry Darwinian natural selection with Mendelian genetics, has also been criticized for its inability to explain macroevolutionary changes—significant changes that occur at or above the level of species. Critics point out that while the Modern Synthesis may account for small-scale changes (microevolution), it struggles to provide satisfactory explanations for large-scale changes (macroevolution) that lead to the emergence of new species, genera, or even higher taxonomic groups.

In summary, while evolutionary theory has provided valuable frameworks for understanding biological diversity and change, it is not without its limitations and criticisms. These skeptical viewpoints serve as a reminder that even widely accepted scientific theories should be subject to ongoing scrutiny and debate. By examining these criticisms, we can aim for a more nuanced and robust understanding of evolutionary processes.

Critical Perspectives from Scientists Who Have Critiqued the General Concept of Evolutionary Theory

To further substantiate the skeptical viewpoints on evolutionary theory, it's essential to consider the criticisms from reputable scientists in the field. Here are some notable quotes:

Stephen Jay Gould, Paleontologist and Evolutionary Biologist:

Gould, known for his theory of punctuated equilibrium, once remarked that the fossil record does not present a smooth and gradual transition of one species evolving into another. While he did not reject evolution, he argued that the standard Darwinian description of gradual change was not fully supported by fossil evidence.

Lynn Margulis, Biologist and University Professor:

Margulis was critical of the neo-Darwinian emphasis on random mutation and natural selection as the sole mechanisms of evolution. She proposed that symbiosis plays a significant role in the evolutionary process, challenging the mainstream view.

Michael Behe, Biochemist and Intelligent Design Advocate:

Behe has been a vocal critic of the Darwinian mechanism of evolution, particularly through his concept of "irreducible complexity." He argues that certain biological systems are too complex to have evolved through a series of small, incremental changes, challenging the mainstream scientific consensus.

Karl Popper, Philosopher of Science:

Popper initially criticized Darwinian evolution for its lack of falsifiability, a key criterion for scientific theories. Although he later revised his stance, his initial criticisms sparked significant debate about the scientific rigor of evolutionary theory.

This highlights that evolutionary theory has faced various criticisms and alternative viewpoints, even from figures within the field of science itself.

The above comments serve to illustrate that skepticism about evolutionary theory is not limited to religious or anti-science perspectives but includes criticisms from within the scientific community itself. Such critiques invite us to consider the limitations and assumptions of evolutionary theory critically.

Section 2

Key Principles
and Mechanisms

Introduction to the Mechanisms that Drive Evolution

Evolutionary theory posits that life on Earth has evolved over billions of years through a series of natural processes. The most commonly cited mechanisms that drive this evolutionary change are natural selection, mutation, genetic drift, and gene flow. These mechanisms are thought to work in concert to shape the genetic makeup of populations over time, leading to the emergence of new species and the extinction of others.

Natural Selection: A Closer Examination

Natural selection is a cornerstone concept in the realm of evolutionary biology, a theory that gained widespread attention and acceptance largely due to the seminal work of Charles Darwin. Often encapsulated by the phrase "survival of the fittest," this mechanism posits that organisms possessing traits advantageous for survival within their specific environmental context are more likely to live long enough to reproduce. Consequently, these advantageous traits are passed down to subsequent generations, gradually becoming more prevalent within the population over extended periods of time.

In this framework, the term "fitness" is not merely a measure of an organism's physical prowess or strength; rather, it encompasses a broader range of attributes, including the ability to find food, evade predators, and successfully mate. These attributes collectively contribute to an organism's overall likelihood of surviving and reproducing in its given environment. As such, natural selection operates on the principle that those organisms better "fit" to their environment—those with traits that confer specific survival advantages—are more likely to propagate their genetic material, thereby influencing the genetic makeup of future generations.

Over successive generations, this process is believed to lead to the gradual accumulation of these advantageous traits within the population, potentially giving rise to new species entirely. This is the crux of how natural selection is thought to drive evolutionary change, serving as a natural "filter" that sifts through genetic variations to promote those that are most conducive to survival and reproduction in a given ecological setting.

Mutation

Mutation is the process by which new genetic variations are introduced into a population. These can range from small changes in a single nucleotide to large-scale alterations involving entire chromosomes. While most mutations are neutral or harmful, a rare few may confer an advantage in a specific environment, providing the raw material upon which natural selection can act.

Genetic Drift

Genetic drift refers to random changes in the frequency of alleles within a population. Unlike natural selection, which is a non-random process, genetic drift does not necessarily lead to traits that are advantageous or disadvantageous. Over time, especially in small populations, genetic drift can lead to the loss of genetic diversity.

Gene Flow

Gene flow is the transfer of genetic material from one population to another, typically through migration or the exchange of gametes. This mechanism serves to increase genetic diversity within populations and can introduce new traits that may be subject to natural selection.

A Skeptical Examination of Key Principles and Mechanisms in Evolutionary Theory

While the concept of natural selection has been widely accepted and propagated as a fundamental mechanism driving evolutionary change, it is crucial to approach this idea with a degree of skepticism. One of the primary criticisms levied against the principle of natural selection is its apparent tautological nature. Critics argue that the phrase "survival of the fittest" is essentially a circular argument, as "fitness" is often defined by survival and reproductive success. In this light, the statement could be rephrased as "survival of those who survive," which doesn't offer much in the way of explanatory power.

Moreover, the concept of "fitness" itself is fraught with ambiguity. While it is generally understood to encompass an organism's ability to survive and reproduce, measuring this "fitness" in real-world scenarios can be exceedingly complex.

Factors such as environmental changes, interactions with other species, and random events can all impact an organism's survival, making it difficult to isolate the role of specific traits in contributing to "fitness."

Another point of contention is the assumption that natural selection operates in a gradual, linear fashion to produce complex biological features. Critics argue that this gradualist perspective fails to account for the "irreducible complexity" observed in biological systems—complex features that necessitate the simultaneous presence of multiple interacting parts to function. In such cases, it is challenging to envision how these features could have evolved incrementally, as the absence of even one component would render the entire system nonfunctional.

Additionally, the role of random genetic mutations as a source of new traits for natural selection to act upon is a subject of ongoing debate. While mutations are generally considered the raw material for evolutionary change, the vast majority of mutations are either neutral or deleterious in terms of their impact on an organism's fitness. The likelihood of a random mutation leading to a significantly advantageous trait is exceedingly low, raising questions about the sufficiency of mutations as a driving force for complex evolutionary changes.

Furthermore, natural selection is often presented as a universal explanation for a wide array of biological phenomena, from the development of antibiotic resistance in bacteria to the intricate structures of the human eye. However, critics argue that the explanatory scope of natural selection is not as broad as often claimed. There are numerous biological features and behaviors that do not appear to offer any survival advantage, yet they persist in populations, challenging the notion that natural selection accounts for all forms of biological diversity.

In summary, while natural selection and other key principles of evolutionary theory offer compelling frameworks for understanding the diversity of life, it is essential to scrutinize these ideas critically. Various aspects of these mechanisms, from their definitional clarity to their explanatory scope, have been called into question, inviting further investigation and debate within the scientific community.

Historical Context

Discussion on the Historical Development of Evolutionary Theory

The historical development of evolutionary theory is a complex tapestry that weaves together the contributions of numerous scientists, naturalists, and thinkers, each adding their own threads to the intricate design. While the concept of evolution has ancient roots, it wasn't until the 19th century that it began to take a more formalized shape, largely due to the seminal work of Charles Darwin and Alfred Russel Wallace.

The idea that species change over time can be traced back to early Greek philosophers like Anaximander and Empedocles, who posited that life originated in the sea and gradually adapted to terrestrial environments. However, these early musings were more speculative than empirical, lacking the rigorous scientific framework that would later be established.

Fast forward to the 19th century, a period that can be considered the crucible in which modern evolutionary theory was forged. Charles Darwin, an English naturalist, set sail on the HMS

Beagle in 1831 on a voyage that would take him to the Galápagos Islands, among other places. It was here that he observed the variations in finches and tortoises from island to island, which sowed the seeds for his groundbreaking theory. Darwin's magnum opus, "On the Origin of Species," published in 1859, introduced the concept of natural selection as the driving force behind evolution. His work laid the foundation for the field of evolutionary biology and forever changed our understanding of the natural world.

Almost simultaneously, Alfred Russel Wallace, another English naturalist, was conducting fieldwork in the Amazon and the Malay Archipelago. Wallace arrived at conclusions remarkably similar to Darwin's about the mechanism of natural selection. The two men presented their findings jointly in 1858, although Darwin is more often credited with the theory due to his more extensive elaboration in "On the Origin of Species."

The late 19th and early 20th centuries saw the incorporation of Mendelian genetics into the framework of evolution, leading to what is known as the "Modern Synthesis." This integrated approach combined the principles of natural selection with the laws of genetic inheritance, providing a more comprehensive explanation for the mechanisms underlying evolutionary change.

In the latter half of the 20th century and into the 21st century, advancements in molecular biology, genomics, and computational biology have further refined and expanded evolutionary theory. These developments have allowed scientists to explore evolutionary processes at the genetic and molecular levels, adding another layer of complexity and understanding to the theory.

However, it's crucial to note that the historical development of evolutionary theory has not been a smooth, linear progression. It has been punctuated by debates, controversies, and paradigm shifts, as new evidence has emerged and as scientists have sought to reconcile various aspects of the theory with new discoveries. This makes the history of evolutionary theory not just a chronicle of accumulating knowledge, but also a case study in the complexities and nuances of scientific progress.

Key Figures in the Development of Evolutionary Theory

The narrative of evolutionary theory is not just a tale of scientific discovery, but also a chronicle of the remarkable individuals who have shaped its course. Among these, two figures stand out for their foundational contributions: Charles Darwin and Alfred Russel Wallace. Their insights and observations laid the groundwork for a new understanding of biological diversity and the processes that drive it.

Charles Darwin is often the first name that comes to mind when discussing evolution. His extensive observations and meticulous research culminated in the formulation of the theory of natural selection. Darwin's journey aboard the HMS Beagle provided him with a wealth of empirical data, as he studied the rich variety of life and the subtle differences between species and varieties on different islands. His careful analysis of finch beak variations and tortoise shell shapes led him to propose that species are not immutable but evolve over time through natural selection, where those best adapted to their environment survive and reproduce.

Alfred Russel Wallace is sometimes overshadowed by Darwin, yet his role in the development of evolutionary theory is no less significant. Wallace's independent conception of natural selection came about through his own extensive fieldwork in the Amazon River basin and the Malay Archipelago. His observations on the geographic distribution of species contributed to the emerging understanding of biogeography and provided strong evidence for the process of evolution.

Beyond these two giants, there are others whose work has been instrumental in the advancement of evolutionary thought. Gregor Mendel, the father of genetics, although not directly linked to the theory of evolution during his lifetime, provided the mechanism of inheritance that would later be integrated into the evolutionary framework. His discovery of the laws of inheritance in the late 19th century remained largely unknown until they were rediscovered at the turn of the 20th century, leading to the synthesis of genetics and evolution.

The Modern Synthesis of the early 20th century brought together a number of key figures, including Theodosius Dobzhansky, Ernst Mayr, and Julian Huxley. Dobzhansky's work on the genetic diversity of populations was pivotal in understanding how evolutionary processes occur at the genetic level. Ernst Mayr's contributions to the concept of species and speciation were fundamental in clarifying how new species arise. Julian Huxley, a proponent of Darwin's work, was instrumental in popularizing the Modern Synthesis, which combined Mendelian genetics with Darwinian evolution.

In more recent times, figures like Stephen Jay Gould and Richard Dawkins have furthered the discussion of evolutionary theory through their respective contributions. Gould's theory of punctuated equilibrium, co-developed with Niles Eldredge, suggested that evolutionary change occurs in rapid bursts separated by long periods of stability. This theory is considered one of the more controversial ideas in evolutionary biology, primarily because it challenges the traditional view of gradualism that was central to Darwin's theory of evolution. Dawkins' gene-centered view of evolution emphasized the role of genes as the primary units of selection.

Skeptical View on Historical Context

The historical context of evolutionary theory is not without its contentious chapters, marked by vigorous debates and profound skepticism that have, over time, played a pivotal role in shaping the theory's trajectory. To adopt a skeptical view on this historical context is to delve into the intricate tapestry of scientific discourse that has surrounded evolutionary theory since its inception.

In the mid-19th century, Charles Darwin and Alfred Russel Wallace independently conceived of natural selection as a mechanism for evolution, setting the stage for a paradigm shift in biological science. Darwin's seminal work, "On the Origin of Species," published in 1859, catalyzed a transformation in the understanding of life's diversity, yet it also ignited a firestorm of criticism that has smoldered and occasionally flared up ever since.

Critics of the time, and indeed some contemporaries, questioned the sufficiency of natural selection as a mechanism capable of driving the complexity and variety observed in the natural world. One such skeptic, the renowned biologist Louis Agassiz, remained unconvinced of natural selection's explanatory power, positing instead that new species arose through a series of creative acts.

The skepticism extended beyond the scientific community to the broader public and philosophical realms. Theologians and philosophers grappled with the implications of a naturalistic mechanism for the origin of species, which seemed to undermine teleological explanations of life and encroached upon the domain traditionally reserved for the divine.

As evolutionary theory evolved, incorporating Mendelian genetics into what became known as the Modern Synthesis, the skepticism persisted. Figures like the geneticist Richard Goldschmidt challenged the gradualist view, proposing instead the idea of "hopeful monsters" — significant mutations that could lead to new species in a single bound. This concept, though largely dismissed by the mainstream, underscored the ongoing debate about the pace and nature of evolutionary change.

The introduction of punctuated equilibrium by Stephen Jay Gould and Niles Eldredge in the 1970s further exemplified the dynamic nature of evolutionary theory and the skepticism that accompanies it. This theory suggested that species remain relatively unchanged for long periods, punctuated by brief, intense periods of speciation — a stark contrast to the slow, steady change posited by classical Darwinism.

Even today, the historical context of evolutionary theory is scrutinized by those who question the completeness and accuracy of the fossil record, the mechanisms of speciation, and the very nature of scientific inquiry. Critics argue that the historical narrative of evolution is often presented with an assurance that belies the uncertainties and gaps in knowledge that persist. They advocate for a more nuanced portrayal of evolutionary theory's history, one that acknowledges the contributions of skeptics and critics in shaping and refining the theory over time.

Askeptical view on the historical context of evolutionary theory is not merely a recounting of dissenting voices but a recognition of the essential role that skepticism and debate have played in the scientific process. It is a reminder that scientific theories are not immutable edicts but living documents, subject to revision and refinement in the face of new evidence and perspectives.

Criticisms and debates that arose during the historical development of the theory

The historical development of evolutionary theory is replete with criticisms and debates, each serving as a testament to the vigorous scientific scrutiny that has continuously shaped and tested its foundations. These debates are not mere footnotes of science. They are central narratives that reveal the contentious and dynamic nature of evolutionary thought.

From the outset, Darwin's theory of natural selection faced a barrage of criticism. One of the earliest and most persistent challenges was the absence of a viable mechanism for inheritance. Critics pointed out that without understanding how traits were passed down through generations, the theory of natural selection stood on shaky ground. This criticism held weight until the rediscovery of Gregor Mendel's work on genetics, which provided the missing piece of the puzzle and gave birth to the Modern Synthesis, integrating Darwinian evolution with Mendelian genetics.

Another focal point of debate was the apparent suddenness of new forms appearing in the fossil record, which seemed at odds with the gradual change proposed by Darwin. This led to the formulation of alternative theories, such as Goldschmidt's "hopeful monsters," which posited that major evolutionary leaps could occur in single generational steps. Although this idea was largely dismissed, it underscored the need for a more nuanced understanding of the tempo of evolutionary change.

The introduction of punctuated equilibrium by Gould and Eldredge further fueled the debate by suggesting that evolution is characterized by long periods of stasis interrupted by brief episodes of rapid change. This theory was met with skepticism by those who saw it as a challenge to the Darwinian orthodoxy of gradualism, sparking a debate that would lead to a deeper examination of the fossil record and the processes of speciation.

The role of natural selection itself has been a subject of ongoing debate. While few dispute its importance, some critics have argued that it cannot account for all aspects of evolution. They point to phenomena such as genetic drift, sexual selection, and neutral mutations as factors that can also drive evolutionary change, suggesting a more complex landscape of evolutionary mechanisms than natural selection alone.

Moreover, the concept of "survival of the fittest," often associated with natural selection, has been criticized for its tautological nature — the idea that those who survive are, by definition, the fittest. Critics argue that this circular reasoning does not provide a testable hypothesis and that the theory needs to be articulated in a way that allows for empirical investigation and falsification.

The debates extend beyond the scientific mechanisms to the philosophical implications of evolutionary theory. Some critics have argued that the theory cannot fully explain the emergence of consciousness, morality, and aesthetics, suggesting that there are aspects of human experience that lie beyond the reach of evolutionary explanations.

In sum, the historical development of evolutionary theory has been shaped by a rich tapestry of criticisms and debates. These discussions have not only tested the robustness of the theory but have also led to its refinement and expansion. They underscore the fact that scientific theories are not static but are continually evolving, much like the biological processes they seek to explain. The debates and criticisms are an integral part of the scientific endeavor, driving the pursuit of knowledge and the deepening of our understanding of the natural world.

Critique from scientists and thinkers of evolutionary theory during its formative years

The formative years of evolutionary theory were marked by a diversity of opinions and critiques from prominent scientists and thinkers. Their skepticism often stemmed from the limitations of contemporary scientific knowledge, philosophical beliefs, or alternative interpretations of empirical observations.

Below are various critiques that were levied against evolutionary theory during its nascent stages:

1. Louis Agassiz (1807–1873)

A renowned paleontologist and geologist, Agassiz was a staunch opponent of Darwin's theory. He argued from a standpoint of empirical evidence, claiming that the fossil record did not support the gradual changes predicted by Darwin. Instead, Agassiz observed what he believed to be evidence of sudden appearances.

2. St. George Mivart (1827–1900)

An English biologist who initially supported Darwin's ideas, Mivart later became one of its most articulate critics. In his book "On the Genesis of Species," Mivart raised several objections, including the problem of incipient stages of useful structures. He questioned how natural selection could favor the initial stages of complex structures, like the eye, which would seemingly provide no immediate advantage to the organism.

3. William Paley (1743–1805)

Although Paley's work predates Darwin's "On the Origin of Species," his ideas were often referenced in opposition to evolutionary theory. Paley's argument from design posited that the complexity and functionality of living organisms resembled human-made objects and, therefore, must have been designed by a divine creator. This teleological argument was a cornerstone of the natural theology prevalent at the time.

4. Lord Kelvin (William Thomson, 1824–1907)

A physicist and mathematician, Lord Kelvin criticized evolutionary theory on the basis of thermodynamics and the age of the Earth. He calculated that the Earth was between 20 to 40 million years old, a timespan he believed was insufficient for the gradual evolutionary processes proposed by Darwin to occur.

5. Adam Sedgwick (1785–1873)

A geologist and priest, Sedgwick was one of Darwin's mentors. Despite their friendship, Sedgwick did not accept Darwin's theory, arguing that the fossil record, which he was intimately familiar with, showed evidence of divine intervention and creation rather than gradual evolution.

These critiques reflect the vibrant scientific and philosophical discourse of the time. They were based on the interpretations of available evidence and often highlighted the gaps in the understanding of natural processes that existed before the advent of genetics and modern paleontology. While many of these criticisms have been addressed by subsequent scientific discoveries, they played a crucial role in challenging and refining evolutionary theory, demonstrating the importance of skepticism and critical thinking in the scientific process.

SECTION 3

MODERN SYNTHESIS

The Modern Synthesis represents a pivotal moment in the history of evolutionary biology, a period during which a bridge was constructed between Charles Darwin's theory of natural selection and Gregor Mendel's principles of heredity. This confluence, emerging in the early to mid-20th century, sought to provide a comprehensive framework that could explain the mechanisms of evolution in light of the genetic discoveries of the time.

The architects of the Modern Synthesis, including prominent figures such as Theodosius Dobzhansky, Ernst Mayr, and Julian Huxley, endeavored to reconcile Mendelian genetics with Darwinian evolution. Their efforts culminated in a unified theory that proposed natural selection acted upon genetic variations within a population, variations that were subject to the laws of inheritance as elucidated by Mendel.

This synthesis was not merely a fusion of ideas but a transformative wave that swept across the biological sciences, influencing fields from systematics and paleontology to ecology and behavior. It provided a robust explanatory model for the adaptation and speciation observed in the natural world, and it fortified the theory of evolution with a genetic foundation that could account for the observed diversity of life.

However, as with any scientific theory, the Modern Synthesis is not without its detractors. Critics have pointed out potential oversights and assumptions that may not hold under closer scrutiny. These criticisms range from questioning the sufficiency of natural selection in driving complex evolutionary changes to challenging the notion that all evolutionary phenomena can be neatly explained within the framework of the Synthesis.

In this section, we will explore the foundational elements of the Modern Synthesis, its impact on evolutionary thought, and the ongoing debates that challenge its premises. Through a skeptical lens, we aim to dissect the arguments, weigh the evidence, and consider the voices of those who advocate for a re-evaluation or even a paradigm shift in our understanding of evolution.

Criticisms and Limitations of the Modern Synthesis

The Modern Synthesis, while foundational to contemporary evolutionary biology, has not been immune to criticism. As our understanding of genetics, development, and ecology has deepened, several limitations and oversights of the Synthesis have come to light. These criticisms are not merely academic quibbles but are central to how we conceptualize and investigate evolutionary processes. In this exploration, we will delve into the various critiques that have emerged, providing a factual and comprehensive overview of the debates surrounding the Modern Synthesis.

Genetic Constraints and Developmental Biology

One significant area of critique concerns the role of genetic constraints and the field of developmental biology, often referred to as 'evo-devo'. The Modern Synthesis largely treats the genome as a collection of discrete genes, each with a specific function and effect on phenotype. However, this view is overly simplistic. Genes often have pleiotropic effects, meaning a single gene can influence multiple traits. Furthermore, the regulatory networks that control gene expression are complex and can limit the range of possible phenotypes, a concept not thoroughly accounted for in the Synthesis.

Developmental processes can constrain evolution by limiting the variety of viable phenotypes that can be produced. For example, certain forms may be unviable due to developmental constraints, regardless of their potential adaptive value. This suggests that the raw material upon which natural selection acts is not as infinitely malleable as the Modern Synthesis implies.

The Role of Genetic Drift

The Modern Synthesis emphasizes natural selection as the primary driver of evolutionary change. However, the role of genetic drift—random changes in allele frequencies that occur by chance—has been shown to have a more significant impact, especially in small populations, than the Synthesis originally acknowledged. Genetic drift can lead to the fixation or loss of alleles independent of their adaptive value, challenging the primacy of natural selection as the sole architect of evolutionary change.

Horizontal Gene Transfer and Non-Mendelian Inheritance

The Modern Synthesis is rooted in Mendelian genetics, which posits that traits are inherited in a vertical fashion from parents to offspring. However, the discovery of horizontal gene transfer (HGT)—the movement of genetic material between unrelated organisms—presents a challenge to this view. HGT is particularly common among bacteria and can have profound evolutionary implications, allowing for the rapid acquisition of beneficial genes and contributing to the evolution of antibiotic resistance.

Moreover, non-Mendelian forms of inheritance, such as epigenetic modifications, which can be influenced by environmental factors and in some cases passed on to offspring, are not accounted for in the Synthesis. These epigenetic changes can affect gene expression and phenotype without altering the underlying DNA sequence, suggesting that inheritance is more complex than previously thought.

Macroevolutionary Patterns and Punctuated Equilibria

The Modern Synthesis has also been criticized for its focus on microevolutionary processes while neglecting macroevolutionary patterns. Critics argue that the gradualistic model of change posited by the Synthesis does not adequately explain the rapid bursts of speciation and the stasis observed in the fossil record—a phenomenon described by the theory of punctuated equilibria proposed by Stephen Jay Gould and Niles Eldredge.

Ecological and Environmental Considerations

Lastly, the Synthesis has been critiqued for its lack of ecological context. Evolution does not occur in a vacuum but is deeply influenced by complex ecological interactions and environmental pressures. The Synthesis traditionally underplays the role of these factors in shaping evolutionary trajectories, often treating the environment as a static backdrop rather than a dynamic participant in the evolutionary process.

In summary, while the Modern Synthesis has served as a cornerstone of evolutionary biology, its limitations and the emergence of new data have led to calls for an extended evolutionary synthesis. This would incorporate our expanded understanding of genetics, development, ecology, and other factors that would interact in the rich tapestry of evolution.

Evolutionary Theory in the Modern World

Evolutionary theory, since its inception, has not remained static; it has dynamically interfaced with a multitude of disciplines, adapting and expanding its reach as new scientific frontiers are explored. In the modern context, evolutionary principles have been applied far beyond the borders of biology, influencing fields as diverse as medicine, psychology, and even the realm of digital algorithms. This section aims to unfold the tapestry of evolutionary theory's applications in contemporary science and society, while also casting a critical eye on the limitations and challenges that accompany its extended use.

Interdisciplinary Influence and Application

The tentacles of evolutionary theory have extended into the medical field, where understanding the evolutionary history of pathogens can be crucial for developing treatments and vaccines. The concept of antibiotic resistance, for instance, is a direct application of the theory, illustrating how microorganisms evolve in response to selective pressures imposed by pharmaceutical interventions. Similarly, the field of psychology has been touched by evolutionary perspectives, with hypotheses like the evolutionary psychology framework attempting to explain human behavior through the lens of adaptation and survival.

In conservation efforts, evolutionary theory informs the management of biodiversity, guiding strategies to preserve genetic diversity and prevent the loss of species. This is predicated on the understanding that evolutionary adaptability is key to the survival of species in changing environments. Moreover, the principles of evolution have found a place in the burgeoning field of artificial intelligence, particularly in the development of evolutionary algorithms that simulate the process of natural selection to optimize problem-solving strategies.

Contemporary Criticisms and Challenges

Despite the widespread application of evolutionary theory, its extension into various domains has not been without criticism. In medicine, the rapid pace of pathogen evolution sometimes outstrips the development of new treatments, leading to questions about the long-term sustainability of current medical approaches to infectious diseases. Critics argue that an over-reliance on evolutionary models may overlook other crucial factors in pathogen-host dynamics.

In the realm of psychology, the application of evolutionary theory has sparked debates about the extent to which complex human behaviors and social structures can be adequately explained by evolutionary processes. Critics point out that the multifaceted nature of human societies and cultures may not be fully accounted for by theories that prioritize genetic and survival imperatives.

Conservation biology faces its own set of challenges, as the application of evolutionary principles to ecosystem management must contend with the unpredictable nature of environmental changes and human impacts. Critics caution against a simplistic application of evolutionary theory, advocating for a more nuanced approach that considers ecological, social, and ethical dimensions.

Lastly, the use of evolutionary principles in artificial intelligence raises philosophical questions about the nature of intelligence and the ethics of creating self-evolving systems. Critics question whether evolutionary algorithms truly mimic natural processes or if they are merely a computational convenience that glosses over the complexity of biological evolution.

CHAPTER 2

DATING METHODS

Introduction

In the grand narrative of Earth's history, dating methods are the tools by which scientists anchor the past in time. These methodologies are pivotal not only in the study of geological formations but also in tracing the evolutionary lineage of life on our planet. They provide the chronological key to unlock the temporal sequence of Earth's geological and biological events, from the formation of the simplest minerals to the emergence and diversification of complex life forms.

The role of dating methods in evolutionary theory and geology cannot be overstated. In evolutionary biology, they allow us to place the emergence of new species and the extinction of others within a temporal context, offering insights into the pace and rhythm of evolutionary change. In geology, these methods help us understand the timing and duration of events such as volcanic eruptions, the formation of mountain ranges, and the shifting of continents. Without the ability to date rocks and fossils, the story of our planet would remain a narrative without a timeline, full of events but devoid of the when.

This chapter is dedicated to a meticulous and critical examination of these dating methods. Our objective is not to undermine their value but to scrutinize their foundations, applications, and the controversies they engender. We will dissect the assumptions upon which these methods are built, the interpretations they lead to, and the debates they continue to fuel. Through this exploration, we aim to present a comprehensive view that acknowledges both the strengths and the limitations of these scientific tools, thereby fostering a more nuanced understanding of the temporal scaffolding that supports much of our knowledge of Earth's past.

With a critical eye, we will navigate through the various techniques, from the well-known radiometric and carbon dating to the less commonly discussed methods like luminescence and dendrochronology. Each method will be evaluated for its contribution to the scientific narrative as well as its potential pitfalls and the skepticism it has attracted. In doing so, we will illuminate the intricate interplay between time, science, and the quest for knowledge in the ever-evolving landscape of Earth's history.

SECTION 1

RADIOMETRIC DATING

Explanation of Radiometric Dating

At the heart of understanding the age of Earth's materials is radiometric dating, a method that relies on the principles of radioactive decay. It is a technique that allows scientists to determine how many years have passed since a rock or a mineral was formed. To comprehend this process, one must first grasp the concept of isotopes, which are variants of chemical elements that have the same number of protons but a different number of neutrons in their nuclei. Some isotopes are stable, but others are unstable or radioactive, meaning they decay over time into other elements in a predictable fashion.

Radiometric dating hinges on the fact that radioactive isotopes decay at a constant, known rate, expressed as a half-life—the time it takes for half of the original radioactive isotope to transform into another element. By measuring the ratio of the parent isotope to the daughter product, and knowing the half-life, the age of the sample can be calculated. This is akin to a cosmic clock, ticking away the eons, with isotopes as the clock's hands, moving at a pace dictated by their half-life.

The process begins when a rock solidifies from magma or lava, a moment when the radioactive "clock" is effectively set to zero. As time passes, the unstable radioactive parent isotopes decay into stable daughter isotopes. By examining the proportions of these isotopes in a rock sample, geologists can back-calculate to determine the time of the rock's formation. This method is not a simple one-off measurement but involves rigorous analytical procedures to ensure accuracy and reliability.

Radiometric dating is not a singular technique but a suite of methods, each based on different radioactive isotopes with varying half-lives, suitable for dating materials of different ages and origins. For instance, uranium-lead dating is used to date ancient rocks billions of years old, while carbon-14 dating is suitable for dating organic remains up to about 50,000 years old.

The reliability of radiometric dating is bolstered by cross-verification. Different methods can be applied to the same rock, and when they yield the same age, the confidence in the date increases. Moreover, radiometric dates can be correlated with other dating methods, such as dendrochronology or ice core dating, to build a cohesive and corroborated geological timeline.

However, it is definitely not without its challenges. The precision of radiometric dating can be affected by factors such as the initial composition of the rock, possible contamination, and the assumption that the decay rate remains constant over immense periods. These are the areas where skepticism finds fertile ground, prompting ongoing research and debate.

In the following subsections, we will delve deeper into the types of radiometric dating, their specific applications, and the skeptical viewpoints that challenge the orthodox interpretations of these scientific timekeepers.

Types of Radiometric Dating: Uranium-Lead, Potassium-Argon, etc.

Radiometric dating encompasses a variety of techniques, each tailored to specific types of rocks and minerals and spanning different time scales. Here are some of the most widely used methods:

Uranium-Lead (U-Pb) Dating

Uranium-Lead dating, commonly abbreviated as U-Pb dating, is one of the oldest and most refined of the radiometric dating schemes. It can be used to date rocks that formed from about 1 million years to over 4.5 billion years ago with precision in the 0.1–1 percent range. The dating method is based on the radioactive decay of uranium isotopes into stable lead isotopes. Uranium has two primordial isotopes, Uranium-238 and Uranium-235, which decay into Lead-206 and Lead-207 respectively. This decay process offers a significant advantage for geochronology because the two uranium isotopes decay at different rates.

The U-Pb dating technique relies on the fact that uranium is a radioactive element that naturally occurs in crystals in trace amounts. Over time, uranium decays following a series of alpha (and beta) decays, in which 238U and 235U decay to stable isotopes of lead (206Pb and 207Pb respectively). This decay occurs at a known rate, characterized by the half-lives of the respective decay processes: 4.47 billion years for 238U to 206Pb, and 704 million years for 235U to 207Pb.

The U-Pb dating method uses the ratio of the parent 238U to the daughter 206Pb and the ratio of the parent 235U to the daughter 207Pb to determine the age of the mineral or rock. Because the half-lives are different, the 238U/206Pb and 235U/207Pb ratios will increase over time, but at different rates. By comparing the amount of uranium to the amount of lead in a sample, the age of the sample can be determined. This process is complicated by the fact that each of the two uranium isotopes decays to a different lead isotope, and this difference in decay rates can be used to provide additional information that can be used to correct for differences in the initial starting conditions.

One of the key strengths of the U-Pb system is that it preserves a record of the original isotopic composition of the mineral at the time of its formation. Zircon, a robust mineral that is very resistant to chemical weathering and thermal alteration, is particularly useful for U-Pb dating. Zircon often contains trace amounts of uranium but no or negligible initial lead. Thus, we can assume that all the lead in the zircon has been produced from the decay of uranium. When zircon forms in igneous rocks, the crystals readily incorporate uranium but reject lead. Therefore, if we measure the ratio of U to Pb in a zircon crystal, we can tell how long it has been since the crystal formed, which is the same as saying we can tell how long it's been since the igneous rock that contains the zircon crystal cooled from a magma.

The U-Pb dating method can be complicated by the presence of inherited lead, lead loss, uranium leaching, and other factors. To address these issues, geologists often use a technique called isochron dating, which involves plotting the ratios of one lead isotope to another on a graph, which can reveal if the sample has been subject to disturbance since its formation. If the data points on the graph are colinear, it suggests the system has remained closed, and the slope of the line gives the age of the rock.

In summary, U-Pb dating is a powerful and sophisticated tool that geologists use to determine the ages of some of the oldest rocks on Earth. It is particularly useful for dating zircon crystals within igneous rocks. However, like all dating methods, it is subject to certain assumptions and potential errors, and interpreting U-Pb ages can be complex, requiring a detailed understanding of the geological history of the area being studied.

Potassium-Argon (K-Ar) Dating

Potassium-Argon dating is a radiometric technique that is used to date rocks and minerals based on the radioactive decay of an isotope of potassium (K) into argon (Ar), a noble gas. Potassium-40 (^{40}K) is a radioactive isotope of potassium that decays into two different products: calcium-40 (^{40}Ca) through beta decay, and argon-40 (^{40}Ar) through electron capture or positron emission. The decay of potassium-40 to argon-40 is of particular interest in dating geological materials because argon, being a gas, can escape from molten rock. This means that any argon that is found in a crystal likely formed as a result of the decay of potassium after the rock solidified.

The K-Ar dating method is based on measuring the accumulation of argon as a by-product of the radioactive decay of potassium within a sample. Because the half-life of potassium-40 is relatively long (about 1.25 billion years), it allows for the dating of materials that are billions of years old. The key to this method is that when volcanic rocks are heated to extremely high temperatures, they release any argon gas trapped in them. As the rock cools and solidifies, it starts to accumulate argon produced by the radioactive decay of potassium. By comparing the proportion of K-40 to Ar-40 in a sample of volcanic rock, and if knowing the decay rate of K-40, the date that the rock formed could be estimated.

Rubidium-Strontium (Rb-Sr) Dating

Rubidium-Strontium dating is another radiometric dating technique, which is used to date rocks and minerals based on the decay of rubidium-87 (^{87}Rb) to strontium-87 (^{87}Sr). Rubidium is an element that is found in many common rock-forming minerals in which it substitutes for the major element potassium. Because rubidium is concentrated in crustal rocks, the continents have a much higher abundance of the isotope compared to the underlying mantle. Strontium, on the other hand, is an element that forms naturally in several isotopic forms, one of which, strontium-87, is formed as a result of the radioactive decay of rubidium-87.

The Rb-Sr dating method is based on the radioactive decay of ^{87}Rb to ^{87}Sr. Rubidium-87 has a half-life of approximately 49 billion years, which is considerably longer than that of potassium-40, making the Rb-Sr dating method useful for dating some of the oldest rocks on Earth. The method works on the principle that the rubidium-strontium isotope pair is immune to chemical reworking because the isotopes are professional tracers of igneous rock differentiation. As a result, once the rock has cooled and solidified, the isotopes do not exchange with the environment.

Carbon-14 (Radiocarbon) Dating

Carbon-14 dating, also known as radiocarbon dating, is a method of determining the age of an object containing organic material by using the properties of radiocarbon, a radioactive isotope of carbon. The technique was developed by Willard Libby and his colleagues at the University of Chicago in 1949, earning Libby the Nobel Prize in Chemistry for his work in 1960.

The method is based on the fact that radiocarbon (14C) is constantly being created in the atmosphere by the interaction of cosmic rays with atmospheric nitrogen. The resulting 14C combines with atmospheric oxygen to form radioactive carbon dioxide, which is incorporated into plants by photosynthesis. Animals then acquire 14C by eating the plants. When the plant or animal dies, it stops exchanging carbon with its environment, and from that point onwards, the amount of 14C it contains begins to decrease as the 14C undergoes radioactive decay. Measuring the amount of 14C in a sample from a dead plant or animal, such as a piece of wood or a fragment of bone, provides information that can be used to calculate when the animal or plant died.

The half-life of 14C is about 5,730 years, which makes it only reliable for dating fossils up to about 50,000 years old. Fossils older than that contain too little 14C to be dated in this way. Therefore, radiocarbon dating is not used to date rocks, which can be millions or billions of years old.

To perform radiocarbon dating, scientists convert the sample to a form they can measure in a mass spectrometer or with liquid scintillation counting, two techniques that measure the amount of radioactive decay to determine the age of the sample. The precision of the dating is enhanced by measuring the ratio of 14C to stable carbon isotopes (12C and 13C), which provides a measure of the concentration of 14C in the sample at the time of death.

Despite its widespread application and general reliability, radiocarbon dating does have limitations. For instance, it assumes that the production rate of 14C by cosmic rays has remained constant over time, which is not always the case. Solar flares, magnetic field variations, and changes in the Earth's atmosphere can all affect the production rate of 14C. Additionally, the method assumes that the ratio of 14C to 12C has remained constant in the atmosphere over time, which is also subject to variation. Calibration curves have been developed based on tree ring data and other sources to correct for these variations, but these corrections are not always perfect.

Contamination is another issue, as the introduction of "modern" carbon during the sample preparation process can result in a sample being dated as significantly younger than its true age. Conversely, older carbon, such as coal or petroleum derivatives, can cause a sample to appear older than it is. Scientists must carefully avoid contamination and use various methods to clean and prepare samples to mitigate these issues.

While radiocarbon dating is a tool widely used by archaeologists, paleontologists, and others, it is not without its challenges and limitations. These must be carefully considered when interpreting the results of radiocarbon dating tests.

Thorium-232 Dating

This method involves the decay of thorium-232 into lead-208 and is used to date corals and marine sediments. It can be particularly useful in conjunction with other dating methods.

Each of these methods requires careful sample preparation and analysis to ensure that the measurements are accurate. The techniques often involve mass spectrometry or other sophisticated instrumentation to measure the isotopic ratios with high precision.

Despite the wide use of these methods, they are not immune to criticism. Skeptics point to potential issues with initial isotopic compositions, contamination, and the assumption of constant decay rates over geological time. These critiques are important as they drive the scientific community to refine their methods and understanding continually.

Skeptical View on Radiometric Dating

Issues with Assumptions Behind Radiometric Dating

Radiometric dating is underpinned by several fundamental assumptions that, if incorrect, can lead to significant inaccuracies. One of the primary assumptions is that the decay rates of isotopes are constant over time. However, some researchers argue that there could be potential influences from environmental factors or other forces that could alter these rates. Another assumption is the initial condition of the rock sample; radiometric dating presupposes that no parent or daughter isotopes were added or removed from the rock since its formation. This assumption is particularly contentious, as it is nearly impossible to certify the original composition of a rock.

Cases Where Radiometric Dating Has Provided Inconsistent or Inaccurate Results

There have been notable instances where radiometric dating has yielded results that are at odds with the established geological timeline. For example, rocks from recently erupted volcanoes have been dated to be millions of years old, and samples from different dating methods often give widely varying ages. Critics argue that these discrepancies cast doubt on the reliability of radiometric dating as a whole.

Comments from Scientists Who Have Critiqued Radiometric Dating Methods

Several scientists have voiced their concerns regarding the reliability of radiometric dating. For instance, Dr. Robert L. Whitelaw, was a nuclear engineer who argued against the reliability of radiometric dating, particularly carbon-14 dating.

Dr. John Baumgardner, a geophysicist, has raised concerns about the assumption of closed systems in radiometric dating. He states, "The U-Pb and Pb-Pb methods routinely give discordant dates and often indicate evidence of open system behavior."

Dr. Steven A. Austin, a geologist, has critiqued the interpretation of isochron data, which is often used in radiometric dating. He suggests that "Isochrons are often interpreted to give a date, but the date does not have a simple meaning because it can represent a mixture of ages."

Dr. Don DeYoung, a physicist and professor at Grace College in Indiana, has questioned the consistency of radiometric dating, particularly when different methods are applied to the same rock sample.

Dr. Jonathan Sarfati, a physical chemist, has criticized the selective reporting of dates. He points out that "Often, only the dates that fit the geological model are published, and those that do not align with preconceived notions are dismissed."

Dr. Tas Walker, a geologist and engineer, has highlighted the problem of circular reasoning in radiometric dating, where the method is often calibrated against fossils of known ages based on evolutionary assumptions. He argues, "There is a circularity in reasoning that undermines the objectivity of radiometric dating."

These comments reflect a broader skepticism within certain scientific circles regarding the assumptions and interpretations of radiometric dating.

SECTION 2

CARBON DATING

C arbon dating, formally known as radiocarbon dating, is a scientific method utilized for determining the age of an object that contains organic material by measuring the levels of carbon-14 (C-14), a naturally occurring radioactive isotope of carbon. This technique hinges on a critical aspect of atmospheric science: the constant formation of C-14 through the interaction of cosmic rays—high-energy particles from outer space—with the stable isotope of nitrogen, nitrogen-14, in the Earth's atmosphere.

The process begins when cosmic rays collide with atmospheric nitrogen, leading to a reaction that converts nitrogen-14 to carbon-14. This newly formed C-14, unstable and radioactive, then combines with oxygen to create carbon dioxide gas, which contains a mix of both stable and radioactive carbon. This radioactive carbon dioxide is then absorbed by living plants during photosynthesis. Animals, in turn, consume these plants, and thus C-14 is integrated into their bodies as well. As long as an organism is alive, it continues to take in C-14 through respiration or consumption, maintaining a certain level of the isotope within its structure.

Upon the death of the organism, the absorption of C-14 ceases, and the isotope begins to decay at a predictable rate, known as its half-life, which is approximately 5,730 years. This means that after 5,730 years, half of the original amount of C-14 in the deceased organism will have decayed into the stable nitrogen-14. By measuring the ratio of C-14 to the stable carbon isotopes, C-12 and C-13, in a sample and comparing this with the current ratio in the atmosphere, scientists can estimate the time that has elapsed since the death of the organism. This method is a cornerstone in the fields of archaeology, geology, and environmental science for dating artifacts, geological strata, and remnants of past life.

However, the precision of this method is contingent upon the assumption that the level of C-14 in the atmosphere has remained constant over the ages. To enhance accuracy, scientists have developed calibration curves that take into account historical fluctuations in atmospheric C-14 levels, determined by analyzing tree rings, ocean sediments, and other proxy data. Despite these measures, the method is most effective for samples that are less than 50,000 years old, beyond which the remaining C-14 levels become too minute to measure reliably.

Applicability and Limitations of Carbon Dating

Carbon dating is applicable primarily to the dating of formerly living things, which have incorporated atmospheric carbon into their biological makeup. This includes a wide array of samples such as wood, charcoal, bone, peat, and carbonate deposits from marine organisms. The method is particularly invaluable in archaeology, where it provides age estimates for artifacts and remains that contribute to our understanding of human history, culture, and environmental changes.

The limitations of carbon dating become apparent when considering the age range and types of materials it can accurately analyze. The effective maximum temporal limit for carbon dating is about 50,000 years. Beyond this point, the residual amount of C-14 is so low that it is difficult to distinguish from background radiation and cosmic rays that may affect the sample. This makes it unsuitable for dating geological formations or fossils that are millions of years old, which require other forms of radiometric dating.

Another limitation is the assumption that the production of C-14 in the atmosphere has been relatively constant over time. However, solar activity, geomagnetic field strength, and changes in the Earth's atmosphere can affect the production rate of C-14. To address this, calibration curves have been developed using data from other dating methods such as dendrochronology (tree-ring dating), which provide a more accurate measure of time by correcting for these variations.

Contamination is also a significant concern. If a sample has been contaminated by additional carbon from fossil fuels or other sources, it can appear much younger than it actually is. Similarly, if it has been exposed to a significant amount of atmospheric C-14 post-mortem, it may seem older. Therefore, careful preparation and pretreatment of samples are crucial to remove any potential contaminants.

Lastly, the precision of carbon dating can be affected by the sample size. Smaller samples may have less C-14 to begin with, making the decay and subsequent measurements less reliable.

Advances in accelerator mass spectrometry (AMS) have mitigated this issue somewhat by allowing accurate measurements of smaller samples. Carbon method and its applicability is bounded by temporal, material, and environmental constraints that must be carefully considered when interpreting the results.

Skeptical View on Carbon Dating: Problems such as Contamination and Calibration

The reliability of carbon dating has been a subject of debate, with skeptics pointing to various problems that can arise, particularly concerning contamination and calibration. Contamination refers to the introduction of foreign carbon into a sample, which can occur during its burial, excavation, or handling. This extraneous carbon can significantly skew the results, leading to inaccurate age estimates. For instance, a piece of ancient charcoal handled with bare hands might absorb modern carbon, making it appear younger than it truly is.

Calibration also presents challenges. Carbon dating relies on the assumption that the C-14/C-12 ratio in the atmosphere has remained constant over time. However, this ratio is known to fluctuate due to various factors, including solar activity, volcanic eruptions, and human activity such as the burning of fossil fuels. These fluctuations can lead to discrepancies between radiocarbon years and calendar years. To mitigate this, calibration curves have been developed, but they are based on historical data and assumptions that may not hold true for all periods and locations.

Furthermore, the calibration process itself can introduce uncertainties. It often relies on cross-referencing C-14 data with other dating methods, such as dendrochronology, which is not without its own set of assumptions and potential errors. The further back the calibration goes, the fewer data points there are to anchor the curve, increasing the margin of error.

Skeptics of carbon dating also highlight cases where known historical dates conflict with radiocarbon dates, even after calibration. These discrepancies can sometimes be significant and have led to re-evaluations of historical timelines. For example, the dating of the eruption of Thera (Santorini) has been a point of contention, with radiocarbon dating suggesting a date different from that derived from archaeological evidence.

Cases Where Carbon Dating Has Been Called into Question

Carbon dating, while a revolutionary tool in archaeology and geology, has faced scrutiny over instances where its results have been at odds with other evidence. One of the most prominent cases is the dating of the Shroud of Turin. Radiocarbon analysis in 1988 dated the shroud to the medieval period, yet various arguments regarding potential contamination and the presence of a biofilm on the fibers have led some researchers to question the validity of these results.

Another notable example is the dating of the Dead Sea Scrolls. Initial carbon dating placed the scrolls in a time frame consistent with the historical and paleographic evidence. However, later tests on different fragments yielded varying ages, leading to debates about the scrolls' authenticity and the reliability of carbon dating, especially when dealing with ancient textiles that may have been exposed to various contamination sources over millennia.

The marine reservoir effect also complicates carbon dating, particularly for organisms that live in the ocean and for humans who consume significant amounts of seafood. This effect occurs because dissolved carbon in the oceans may have a different radiocarbon age than the atmosphere, due to the upwelling of deep, ancient waters. Consequently, marine organisms can appear to be hundreds or even thousands of years old when they are actually much younger. Studies have shown that without proper correction, this effect can lead to significant dating inaccuracies, which has implications for the dating of maritime cultures and coastal settlements.

These cases highlight the complexities and potential pitfalls of carbon dating, emphasizing the need for cross-verification with other dating methods and for a cautious interpretation of radiocarbon data. Critics argue that these examples underscore the limitations of the method and advocate for a more skeptical approach to its results, especially when they conflict with other types of evidence.

The Marine Reservoir Effect and Its Impact on Dating Accuracy

The marine reservoir effect describes a discrepancy in radiocarbon ages between marine and terrestrial organisms. This phenomenon occurs because the carbon in ocean water, derived from a mixture of atmospheric CO_2 and ancient, carbon-14-depleted carbon from the deep ocean, has a different isotopic signature than the carbon in the terrestrial biosphere. Marine organisms, which derive their carbon from dissolved bicarbonate in the ocean, inherently contain less carbon-14 than their terrestrial counterparts, which directly assimilate atmospheric CO_2.

The relevance of the marine reservoir effect in radiocarbon dating is profound. It can lead to the misdating of marine-derived samples, which appear older than their true age. This effect varies geographically and through time, influenced by factors such as ocean currents, upwelling of deep-sea waters, and changes in atmospheric CO2 levels. Consequently, samples from individuals with a diet rich in marine foods, or from marine animals themselves, require adjustments to account for this offset.

Understanding and correcting for the marine reservoir effect is crucial for accurate radiocarbon dating.

Description of Carbon-14 Level Differences Between Marine and Atmospheric Environments

The carbon-14 levels in the atmosphere and marine environments are subject to a dynamic equilibrium, yet they are not identical due to several influencing factors. In the atmosphere, carbon-14 is produced by the interaction of cosmic rays with nitrogen atoms, leading to a relatively uniform distribution of this isotope around the globe. Atmospheric carbon dioxide, which contains carbon-14, is then taken up by terrestrial plants during photosynthesis, and subsequently by the animals that consume these plants, leading to a relatively consistent ratio of carbon-14 in terrestrial organisms.

In contrast, the marine environment presents a more complex scenario. The surface waters of the ocean exchange carbon dioxide with the atmosphere, but this exchange is not the sole source of carbon in marine ecosystems. Deep ocean currents circulate carbon from the ocean floor, which is significantly depleted in carbon-14 due to its isolation from the atmosphere for extended periods. This aged carbon mixes with the surface water, diluting the carbon-14 concentration.

Moreover, the rate of carbon-14 incorporation into marine organisms is affected by the solubility of carbon dioxide in seawater, which is influenced by temperature, salinity, and pressure. The result is a marine reservoir of carbon that has a lower concentration of carbon-14 compared to the atmospheric carbon. This discrepancy is further compounded by regional variations, such as upwelling zones where ancient, carbon-14-depleted waters are brought to the surface, creating hotspots of even greater carbon-14 dilution.

These differences in carbon-14 levels between marine and atmospheric environments necessitate careful calibration when dating marine samples. Without accounting for the marine reservoir effect, the radiocarbon dating of marine life or of individuals with a diet heavy in marine foods would yield dates that are erroneously older than their actual age, potentially skewing historical timelines and obscuring our understanding of past events.

Impact of Marine Diet on Radiocarbon Dates

The implications of a marine diet on radiocarbon dating have been highlighted in several studies, particularly those examining human and animal remains. Research has shown that individuals with a high intake of marine foods can exhibit radiocarbon ages that are significantly older than their true age. This is due to the incorporation of "old" carbon from marine sources into their tissues, skewing the radiocarbon dating results.

One notable study examined the remains of individuals from coastal and island communities, where marine diets were prevalent, and found that their radiocarbon dates needed substantial correction to account for the marine reservoir effect. Another study focused on the seals and other marine mammals, which are entirely reliant on marine food webs, demonstrating that their radiocarbon ages could appear hundreds of years older than they actually are.

These findings underscore the importance of considering dietary patterns when interpreting radiocarbon dates. Archaeologists and geologists must account for the marine reservoir effect, especially when working with populations known to consume large amounts of marine resources. Without this consideration, the radiocarbon dating of both human and animal remains could lead to inaccuracies in historical timelines and misunderstandings of past human behaviors and migrations.

The marine reservoir effect represents a formidable challenge in the field of radiocarbon dating, with the potential to skew dates by a substantial margin. In certain instances, particularly in marine environments with fluctuating levels of carbon-14, this effect can lead to inaccuracies in the dating of organisms by up to 2000 years. This discrepancy arises because marine organisms absorb carbon from a source that has a different radiocarbon age than the atmosphere, leading to a significant offset. Such a pronounced deviation necessitates careful calibration and consideration, especially when dating marine life or human populations with a diet heavily reliant on seafood. The implications of this effect are profound, as it can alter historical timelines and necessitate the reevaluation of previously established chronologies.

Case Studies on the Marine Reservoir Effect

The marine reservoir effect has been observed in numerous case studies, providing concrete examples of its impact on radiocarbon dating. These instances serve as cautionary tales for researchers relying on carbon dating for historical and archaeological conclusions.

Here are two notable instances:

The Norse Colonies of Greenland: Radiocarbon dating of human bones from Norse graves in Greenland has shown that a diet rich in marine foods can cause radiocarbon ages to appear several hundred years older than they are. This is due to the marine reservoir effect, where the carbon in the marine food consumed by the Norse had a different radiocarbon age than the contemporary atmosphere.

The Case of the Kennewick Man: The Kennewick Man is a notable archaeological find in North America. Initial radiocarbon dating suggested an age that was later adjusted due to considerations of the marine reservoir effect. The individual's diet, which included a significant amount of marine foods, had to be factored into the dating process to provide a more accurate age estimate.

In addition to these, there have been numerous instances where radiocarbon dating has been challenging or controversial, often leading to significant debates within the scientific community. These debates can arise due to discrepancies between radiocarbon dates and other evidence, or because of the implications of the dates for understanding historical or pre historical events.

These examples highlight the importance of considering the marine reservoir effect when interpreting radiocarbon dating results, especially for individuals or animals with diets high in marine foods.

Significant Issues Posed by the Marine Reservoir Effect

The marine reservoir effect presents a substantial challenge to the accuracy of radiocarbon dating, particularly when dating organisms that have derived a significant portion of their diet from marine sources. The effect arises because the levels of carbon-14 in the marine environment can be significantly different from those in the terrestrial atmosphere. This discrepancy is primarily due to the dilution of surface ocean carbon with carbon from deep ocean waters, which contain lower levels of carbon-14 because they have been isolated from atmospheric carbon for extended periods.

Impact on Calibration Curves

One of the most significant issues is the impact on calibration curves. Radiocarbon dating relies on calibration curves that convert radiocarbon years into calendar years. These curves are based on the assumption that the carbon-14/carbon-12 ratio in the atmosphere is constant over time and across various regions. However, the marine reservoir effect can lead to local, and sometimes regional, variations in this ratio, necessitating the use of separate calibration curves for marine samples.

Variability and Regional Differences

The marine reservoir effect is not uniform across all bodies of water. It varies greatly depending on a multitude of factors, including ocean circulation, upwelling rates, and the proportion of carbon from different sources. This variability can lead to regional differences in the reservoir effect, making it challenging to apply a one-size-fits-all correction for marine samples.

Dietary Influence on Human and Animal Remains

For human and animal remains, the marine reservoir effect can lead to inaccuracies in radiocarbon dating if the diet of the individuals was rich in marine food sources. This is particularly problematic for coastal or island populations, where seafood might constitute a significant part of the diet. Without proper correction, this can result in dates that are erroneously older or younger than they actually are.

Implications for Historical and Archaeological Contexts

The marine reservoir effect has profound implications for historical and archaeological contexts. It can affect the dating of marine sediments, shell mounds, human remains, and any artifacts associated with marine food consumption. This can lead to misinterpretations of the timing and sequence of events in human history, such as migrations, trade, and the development of certain technologies or practices.

To address the issues posed by the marine reservoir effect, researchers must try to develop localized correction factors. This involves measuring the reservoir effect in various regions and for different species, which can then be used to adjust radiocarbon ages. Such measures are critical for ensuring that the radiocarbon dating of marine-influenced samples yields results that are as accurate as possible.

SECTION 3

DENDROCHRONOLOGY
(TREE-RING DATING)

Explanation of dendrochronology

Dendrochronology, commonly known as tree-ring dating, is a method of dating based on the analysis of patterns of tree rings, also known as growth rings. Each ring would typically mark the passage of one year in the life of the tree. This method is used to date the exact year a tree was felled, to study changes in the environment such as climate change, and to calibrate radiocarbon ages.

The science of dendrochronology is based on the phenomenon that in temperate climates, trees generally add one growth ring per year. These rings are visible as concentric circles on cross-sections of the trunk. The width of a tree ring shows the amount of growth that has occurred during one growing season and can provide a clue to the environmental conditions that prevailed during that time. For instance, a wide ring could indicate a year of plentiful rainfall or optimal growth conditions, while a narrow ring could suggest a year of drought or other stress on the tree.

Dendrochronologists, scientists who study tree rings, examine these ring patterns and match them with the patterns from trees of the same species and age. By building a chronological sequence from living trees back to ancient wood, they try to create a continuous record that spans many centuries. This record can then be used to date wooden objects, such as buildings or artifacts, and to provide information about past climates.

When used in conjunction with other dating methods, such as radiocarbon dating, dendrochronology can provide a calibration tool that corrects for variations in the concentration of carbon-14 in the atmosphere over time. This calibration is essential for achieving accurate radiocarbon dating results, especially when dating objects that are several thousand years old.

The method of dendrochronology makes it an invaluable tool for archaeologists, geologists, and historians.

How Dendrochronology is Used in Conjunction with Other Dating Methods

Dendrochronology is often used in conjunction with other dating methods to calibrate and validate them. For instance, radiocarbon dating can be cross-referenced with tree-ring dates to correct for variations in atmospheric C-14 over time. This process, known as calibration, is crucial for obtaining accurate radiocarbon dates and involves matching radiocarbon dates with corresponding dendrochronological dates.

In archaeology, dendrochronology can provide precise dates for wooden structures or artifacts, offering a timeline that can be used to cross-date other materials found at the site. It is particularly useful in dating the wood panels of paintings or wooden artifacts whose age is unknown.

In geological studies, tree rings can be used to date events such as volcanic eruptions or glacial advances by examining the growth interruptions or anomalies in the rings that correspond to these events. These dendrochronological dates can then be used to calibrate other geological dating methods.

Furthermore, dendrochronology is used in the calibration of thermoluminescence dating, which is often applied to ceramics, and optically stimulated luminescence dating, used for sediments. By providing a known-age sample from tree rings, these methods can be fine-tuned to yield more accurate results.

The integration of dendrochronology with other dating methods is a powerful tool in the scientific dating arsenal, allowing for the cross-verification of dates and the construction of a more robust and detailed chronology of historical and geological events.

Skeptical View on Dendrochronology: Limitations and Criticisms

Dendrochronology, while a powerful and often precise dating method, is not without its limitations and has been subject to various criticisms. Skeptics of dendrochronology point out several potential issues that may affect the accuracy and reliability of tree-ring dating.

Inherent Limitations of Tree Growth

Dendrochronology relies on the assumption that trees in temperate climates produce a single growth ring each year.

However, this process can be more complex than it seems, leading to potential errors in dating if the inherent limitations of tree growth are not fully understood and taken into account.

False Rings

One of the primary issues is the occurrence of false rings. In certain years, particularly when environmental conditions change dramatically within a single growth season, a tree may produce an extra ring. For example, an unusually warm and wet spring followed by a severe drought might prompt the tree to start a second growth phase once conditions improve, resulting in two rings forming in one year. These false rings can appear indistinguishable from annual growth rings to the untrained eye.

Missing Rings

Conversely, trees may not produce a growth ring in a given year if conditions are especially harsh, such as during extreme droughts, cold spells, or volcanic events that block sunlight. In such cases, the absence of a ring can lead to undercounting the age of the tree and, by extension, misdating the wood in archaeological or geological samples.

Localized Disturbances

Trees may also experience localized disturbances that affect their growth rings. Damage from fire, insects, or disease can disrupt the normal growth pattern of a tree. If such a disturbance is not recognized, it could lead to misinterpretation of the growth rings, as the tree may produce irregular growth in response to the stress.

Growth Suppression and Release

Trees growing in dense forests may experience growth suppression due to competition for light and nutrients. When surrounding trees are removed or die, the surviving trees may exhibit a sudden increase in growth rate, known as a release, which can result in wider rings that might be misinterpreted as indicative of a particularly favorable year.

Anatomical Anomalies

Some trees may exhibit anatomical anomalies in their wood structure that can complicate ring counting. For instance, the presence of intra-annual density fluctuations, which are changes in the density of wood cells within a single growth ring, can be mistaken for multiple rings.

Ring Distortion

The growth rings of a tree may become distorted due to environmental or mechanical stress, such as wind or leaning. This distortion can lead to uneven ring width around the circumference of the tree, potentially causing inaccuracies if the full circumference is not sampled.

Regional and Species-Specific Variations

The interpretation of tree rings is highly dependent on regional climate patterns. Trees from different regions can show different growth patterns even in the same chronological period. Additionally, different species of trees respond to environmental conditions in varying ways. These species-specific responses can complicate the process of cross-dating rings from different trees or different regions.

Human and Environmental Influences

Human activities such as logging, forest management, and environmental changes can influence tree growth. Pollution, climate change, and introduction of non-native species can alter growth patterns, potentially leading to misleading ring patterns. Critics argue that unless these factors are adequately accounted for, dendrochronological data can be skewed.

Limitations in Temporal and Spatial Scope

Dendrochronology is limited to the lifespan of a tree. For very old historical or geological events, living trees are not sufficient, and preserved wood must be used. However, finding well-preserved wood that can be securely dated back to a specific period is often challenging. Furthermore, dendrochronology is primarily applicable in regions where trees with clear annual rings are available, limiting its application in tropical regions where such trees are less common.

Cross-Dating Complexities

Cross-dating, the process of matching ring patterns across different trees, is a cornerstone of dendrochronology. However, this process assumes that the same environmental conditions affect all trees in a region identically. Critics point out that individual trees may have unique responses to environmental factors, leading to discrepancies in ring patterns that can complicate or invalidate cross-dating efforts.

Reliance on Calibration Curves

Tree-ring dating often relies on calibration curves that are constructed from the tree-ring records of known-age wood. Skeptics argue that these curves are based on the assumption that the environmental conditions affecting tree growth in the past are sufficiently understood and that the sample of trees used to construct the curves is representative of the entire population. Any errors in these assumptions can propagate through the calibration process, affecting the accuracy of dates derived from dendrochronology.

Cases of Inconsistencies in Dendrochronology

The Problem of Missing Rings:

In certain instances, especially in trees that have undergone environmental stress, dendrochronology can yield inaccurate results due to missing rings. For example, in the 1960s, researchers studying bristlecone pines in Nevada discovered that some trees were missing rings. This finding was crucial because these trees serve as a baseline for calibrating radiocarbon dating.

The Belfast Oak Chronology:

In Ireland, the Belfast Long Chronology, which was developed using oak tree rings, initially presented discrepancies when compared to radiocarbon dating. It was later found that these discrepancies were due to variations in local climate and environmental conditions affecting tree growth, which were not initially accounted for.

Critiques from Scientists and Thinkers

Dr. Valerie Trouet:

A dendrochronologist herself, Dr. Trouet has pointed out that while tree rings are generally reliable indicators of chronological order, they can be influenced by non-climatic factors such as disease, competition, and human activity, which can lead to erroneous conclusions if not properly accounted for.

Dr. Bernd Becker:

Dr. Becker, another dendrochronologist, has critiqued the use of tree-ring data in isolation. He emphasizes the importance of cross-referencing dendrochronological data with other dating methods and historical records to ensure accuracy.

Dr. Tom Swetnam:

As a specialist in fire history studies through dendrochronology, Dr. Swetnam has highlighted cases where fire scars in tree rings led to underestimation of the actual age of trees, showing that external factors like fire can complicate tree-ring analysis.

Philosopher of Science, Karl Popper:

While not directly related to dendrochronology, Popper's philosophy of science, which emphasizes falsifiability as a criterion for scientific theories, has been applied by some critics to question the reliability of dendrochronology. They argue that dendrochronological hypotheses must be subjected to tests that could potentially falsify them to be considered scientific.

Section 4

Luminescence Dating

Introduction to Luminescence Dating

L uminescence dating represents a collection of scientific methods used to determine the time elapsed since certain mineral grains were last exposed to light or heat. This dating technique is based on the phenomenon of luminescence, which occurs when mineral crystals, such as quartz (hard mineral consisting of silica, found widely in rocks) or feldspar (abundant rock-forming mineral), are stimulated and release previously absorbed energy in the form of light.

The core principle behind luminescence dating lies in the ability of these minerals to trap and store energy from environmental radiation. This energy is absorbed and stored in the form of trapped electrons within defects in the crystal lattice of the minerals. Over time, these electrons accumulate, effectively 'charging' the crystal.

There are two main types of luminescence dating: Optically Stimulated Luminescence (OSL) and Thermoluminescence (TL). Both methods rely on the same fundamental principle but differ in the mechanism used to release the stored energy.

In OSL, the trapped electrons are released by exposing the mineral to light of a specific wavelength. This exposure causes the electrons to become 'unstuck' from the defects in the crystal lattice, recombine with atoms, and emit light, the intensity of which can be measured and used to calculate the time since the mineral was last exposed to light.

Thermoluminescence, on the other hand, involves heating the mineral. The heat causes trapped electrons to gain enough energy to escape from the lattice defects. As these electrons recombine with atoms, they emit light. The amount of light released is proportional to the accumulated radiation dose and, thus, the time that has elapsed since the mineral was last heated.

Both OSL and TL are invaluable in archaeological and geological contexts. They allow scientists to date events such as sediment deposition or the last time a piece of pottery was fired in a kiln, providing crucial insights into human history and geological processes.

Types of Materials Dated Using Luminescence Techniques

Luminescence dating techniques, notably Optically Stimulated Luminescence (OSL) and Thermoluminescence (TL), are versatile in their application, allowing for the dating of a diverse range of materials. The primary materials dated using these methods include sediments and ceramics, each offering unique insights into past environments and human activities.

Sediments:

» Geological Sediments: OSL is particularly useful in dating geological sediments, such as sands and silts. These sediments can be from various environments like river terraces, dunes, and glacial deposits. Understanding the age of these sediments is crucial for reconstructing past landscapes and climate changes.

» Archaeological Sediments: In archaeological contexts, OSL helps date the sediments in which artifacts are found. This provides a timeline for when human activities occurred in a given area, aiding in the reconstruction of human history and migration patterns.

Ceramics:

» Pottery and Fired Clay: TL dating is often used to date ceramics, especially pottery. When clay is fired to make pottery, the intense heat empties the stored luminescent signal. Once cooled, the luminescent signal accumulates again. By measuring the current luminescence level, the last time the pottery was fired can be estimated, which is often indicative of its age and the period of human activity associated with it.

» Brick and Tile: Similar to pottery, bricks and tiles can also be dated using TL. This is especially useful in historical archaeology, where the age of structures can be determined, providing insights into urban development and architectural history.

Other Materials:

» While sediments and ceramics are the most common, luminescence dating can also be applied to other materials like volcanic ash layers and some types of rock surfaces. However, the application to these materials often involves more complex procedures and interpretations.

Luminescence dating, through its application to these materials, provides a window into both the geological past and the chronology of human activities. The accurate dating of sediments offers a timeline of environmental changes, while the dating of ceramics and other human-made materials sheds light on the chronology of human settlements, migrations, and cultural developments.

Optically Stimulated Luminescence (OSL)

Optically Stimulated Luminescence (OSL) is a sophisticated dating technique used to determine the last time certain minerals, primarily quartz or feldspar, were exposed to light. The method is based on the principles of electron trapping and the subsequent release of these electrons through light stimulation, which produces luminescence.

Electron Trapping:

» When minerals like quartz and feldspar are exposed to natural radiation from their surrounding environment, they absorb this energy. This energy excites electrons within the mineral, causing them to become dislodged from their normal positions in the atomic lattice.

» These dislodged electrons then become trapped in defects or imperfections within the mineral's crystal lattice. Over time, the number of trapped electrons accumulates, correlating with the duration of exposure to radiation. This process effectively 'charges' the mineral.

Stimulation and Luminescence:

» In the laboratory, these minerals are stimulated using light at specific wavelengths. This stimulation causes the trapped electrons to gain enough energy to escape their traps.

» As these electrons recombine with their original atoms, they release energy in the form of light, or luminescence. The intensity of this luminescence is proportional to the number of trapped electrons, and hence, the accumulated radiation dose.

Measuring the Age:

» By measuring the intensity of the emitted luminescence, scientists can calculate the accumulated radiation dose, also known as the equivalent dose (D_e).

» The age of the sample is then determined by dividing this equivalent dose by the annual radiation dose (rate of radiation absorption), which is estimated through dosimetry techniques.

Application of OSL in Dating Sediments

The application of Optically Stimulated Luminescence (OSL) in dating sediments is based on its attempt to determine the last time these sediments were exposed to sunlight. While the method is widely used, it's important to approach its results with a degree of skepticism due to the inherent uncertainties and assumptions involved.

Principle of Dating Sediments:

» OSL aims to measure the time elapsed since sediment grains, like quartz or feldspar, were last exposed to sunlight. This exposure to light resets the luminescence 'clock' by emptying the trapped electrons in the mineral grains.

» Once buried, these grains begin to accumulate trapped electrons again due to natural background radiation. The assumption here is that the longer the sediment has been buried, the more trapped electrons it will have accumulated.

Challenges and Limitations:

» A key challenge in using OSL for dating sediments is ensuring that the entire luminescence signal was reset at the time of burial. Partial resetting or incomplete zeroing of the signal can lead to overestimation of age.

» The accuracy of OSL dating also heavily relies on the estimation of the environmental dose rate, which can vary over time. Changes in moisture content, uranium, thorium, and potassium concentrations in the surrounding sediment can affect the dose rate, complicating age calculations.

Archaeological Applications:

» In archaeology, OSL is employed to date the time since artifacts were buried. This is crucial for establishing chronologies of human activities, especially in sites without organic materials suited for radiocarbon dating.

» However, the interpretations from OSL dates must be considered cautiously, particularly in complex stratigraphic contexts where post-depositional processes might have altered the original sediment layers.

Geological Contexts:

» Geologically, OSL is useful in reconstructing paleoenvironmental conditions and understanding sediment deposition histories. It's applied in contexts like fluvial, eolian, and even glacial environments.

» Despite its usefulness, the geological interpretations based on OSL dates should be tempered with an understanding that natural processes like erosion, redeposition, or bioturbation can potentially disrupt the sedimentary layers, affecting the reliability of the dating results.

In both archaeological and geological contexts, while OSL dating provides valuable insights, it's imperative to maintain a skeptical perspective regarding its accuracy and reliability. The method's dependency on multiple assumptions and variable environmental factors necessitates cautious interpretation of the dating results, often requiring corroboration with other dating methods or stratigraphic information.

Understanding Thermoluminescence Dating

Thermoluminescence (TL) dating is a technique that measures the accumulated radiation dose in materials like ceramics, offering a window into the past, particularly in the field of archaeology. This method hinges on the principle that mineral crystals, such as those found in pottery and hearths, trap electrons when exposed to natural background radiation.

The Principle of Electron Trapping and Heating:

» Similar to Optically Stimulated Luminescence (OSL), TL relies on the trapping of electrons within the crystal lattice of minerals when they absorb energy from environmental radiation. Over time, these trapped electrons accumulate, correlating with the total radiation dose the material has absorbed since its last heating.

» The unique aspect of TL is the method of releasing these trapped electrons. Instead of using light, as in OSL, TL uses heat. When a sample, such as a piece of ancient pottery, is heated in a laboratory, the trapped electrons are released from their lattice traps.

Luminescence Upon Heating:

» As the electrons are released during heating, they recombine with atoms in the crystal lattice, emitting photons — this is the thermoluminescence. The amount of light emitted is directly proportional to the number of trapped electrons, and hence, the accumulated radiation dose.

» By measuring this emitted light, archaeologists can calculate the accumulated radiation dose, known as the paleodose.

Dating Ceramics and Cooking Hearths:

» The TL method is particularly valuable in dating ceramics. When pottery is made, the high temperatures of firing effectively empty any previously accumulated luminescent signal in the clay. Once cooled, the pottery begins to accumulate trapped electrons from that point forward.

» By measuring the TL, archaeologists can estimate when the pottery was last fired, which is usually a good indication of its age. This method is also used for dating cooking hearths and other fired archaeological materials, providing an insight into the timing of human activities.

Skeptical Considerations in Thermoluminescence (TL) Dating

While the application of Thermoluminescence (TL) dating has undoubtedly contributed significantly to our understanding of archaeological timelines, especially in the context of dating ceramics and ancient cooking hearths, several critical skeptical considerations must be taken into account. These considerations not only highlight the complexities inherent in TL dating but also underscore the importance of a cautious approach when interpreting its results.

Incomplete Resetting of the Luminescent Signal:

» A fundamental assumption in TL dating is that the heating event (e.g., firing of pottery) completely resets the luminescent signal, erasing all previous accumulations of trapped electrons. However, this assumption may not always hold true. In cases where the firing temperature was insufficient or the duration of heating was inadequate, a partial reset could occur, leaving a residual luminescent signal. This residual signal could lead to an overestimation of age, as it would add to the luminescence emitted during laboratory heating, creating an illusion of a longer period of electron trapping than actually occurred.

Variability in Radiation Dose Rates:

» The accuracy of TL dating hinges on the estimation of the dose rate, which is the rate at which the sample absorbs radiation from its environment. This rate can be influenced by various factors, including the presence of radioactive elements in the surrounding soil or materials, and changes in environmental conditions such as moisture content. Fluctuations in these factors over time can lead to inaccuracies in dose rate calculations, thereby affecting the precision of the dating results. Moreover, the historical radiation environments of the samples are often reconstructed based on current conditions, which may not accurately reflect past environments, further complicating dose rate estimations.

Potential for Laboratory Error:

» The process of measuring TL itself is subject to potential errors. Factors such as the sensitivity of the equipment, the precision in controlling the heating rate, and the calibration of the detectors can all impact the accuracy of the luminescent readings. Inconsistent methodologies or calibration issues across different laboratories can lead to varying results for the same sample, raising questions about the reproducibility and reliability of TL dates.

Interpretational Ambiguities:

» Archaeological contexts often present complex stratigraphy where materials might have been subjected to multiple heating events or relocations. Such complexities can introduce ambiguities in interpreting TL dates. For instance, a piece of pottery might have been used and reheated multiple times before its final discard, complicating the determination of which heating event the TL date actually represents.

Necessity for Corroborative Evidence:

» Given these uncertainties, TL dates are best viewed not as definitive answers but as part of a broader archaeological and geological narrative. Corroborating TL dates with other dating methods or contextual archaeological information is crucial. This multi-faceted approach helps in cross-verifying the findings and building a more robust and reliable historical reconstruction.

In essence, while TL dating is a powerful tool, its application is fraught with potential pitfalls and uncertainties. A skeptical approach, one that critically evaluates the methodology and its underlying assumptions, and seeks corroborative evidence, is essential in ensuring the reliability and accuracy of the dating results.

Implications of Skeptical Considerations on Evolutionary Theory and Archaeology: TL and OSL

The critical considerations of both Thermoluminescence (TL) and Optically Stimulated Luminescence (OSL) dating methods carry significant implications for interpretations in evolutionary theory and archaeology. Understanding these implications is key to appreciating the complexities and challenges in constructing accurate historical and evolutionary narratives.

The potential inaccuracies in TL and OSL, such as incomplete resetting of signals or variability in radiation dose rates, can result in erroneous dating of artifacts and sediments. This misdating can distort our understanding of historical timelines, leading to incorrect assumptions about the age of archaeological sites, the sequence of human activities, or the development of technologies.

For example, an overestimated TL or OSL date for a cooking hearth or a layer of sediment can suggest an earlier human presence in a region than actually occurred, impacting theories related to human migration, settlement patterns, and cultural interactions.

In evolutionary theory, these dating methods are critical for placing key events and developments in an accurate temporal context. Misinterpretations due to dating errors can mislead our understanding of the pace and nature of evolutionary processes.

For instance, the dating of sediment layers containing fossilized remains or artifacts using OSL can significantly influence the perceived timeline of hominid evolution. Similarly, the TL dating of ceramics can reshape our understanding of the technological advancements of early human societies.

The reliance on TL and OSL for dating critical archaeological and geological layers necessitates a rigorous cross-verification process with other dating methods. Discrepancies between different dating techniques can raise questions about the reliability and accuracy of the established timelines.

In archaeological contexts, this might involve correlating TL or OSL dates with radiocarbon dating or stratigraphic analysis, while in geological contexts, it could include comparing these dates with other forms of radiometric dating.

Given these challenges, a multidisciplinary approach that incorporates various dating methods, along with geological, paleoenvironmental, and archaeological analyses, is essential. This approach allows for a more nuanced and robust interpretation of the data, mitigating the limitations inherent in any single dating technique.

The skepticism surrounding TL and OSL dating serves as a reminder of the importance of continuous scientific inquiry and methodological refinement. It encourages researchers to question established narratives and remain open to new interpretations and discoveries.

Case Studies with Controversies in Luminescence Dating

Luminescence dating, encompassing both Thermoluminescence (TL) and Optically Stimulated Luminescence (OSL), has been a pivotal tool in archaeological and geological research. However, its application has not been without controversies and challenges, as highlighted in several notable case studies:

Controversies in Archaeological Dating:

» In various archaeological sites, the application of OSL and TL dating has occasionally led to contentious results. Discrepancies in dating artifacts and structures, such as ancient pottery or habitation layers, have sometimes challenged established historical timelines. Controversies often arise from questions about the complete resetting of the luminescent signal, the accuracy of the estimated radiation dose rates, and the interpretation of the data in complex stratigraphic contexts.

Debates in Geological Dating:

» Geological findings, particularly those involving sediment dating, have also been subject to debate. Luminescence dating has been used to estimate the age of sediment layers, influencing our understanding of geological processes and environmental changes. However, issues such as changes in environmental radiation over time and the potential for post-depositional disturbances have led to debates about the reliability of these dates.

Specific Sites and Findings:

» Certain high-profile sites and findings have become focal points for these debates. For instance, the dating of early human occupation sites in Australia and the Indus Valley Civilization sites in India using luminescence techniques has generated discussion within the scientific community. These cases highlight the complexity of accurately dating ancient materials and the implications of these dates for our understanding of human history and prehistoric environments.

The Evolution of Luminescence Dating:

» Despite these challenges, luminescence dating continues to evolve, with improvements in methodology and a deeper understanding of its limitations. The controversies and debates have spurred further research, leading to more refined and accurate dating techniques, and a more nuanced interpretation of luminescence dating results.

In conclusion, while luminescence dating has significantly advanced our understanding of archaeological and geological histories, it is important to acknowledge the controversies and challenges inherent in these methods. These case studies underscore the need for a cautious and critical approach in interpreting luminescence dating results, reminding us of the ongoing nature of scientific inquiry and discovery.

Conclusion: A Skeptical Review of Dating Methods

As we conclude this chapter , it becomes increasingly clear that while these techniques are invaluable tools in piecing together the history of our planet and the story of human evolution, they are definitely not without their limitations and are areas of heated debate.

The exploration of radiometric and carbon dating methods revealed the complexities and assumptions inherent in measuring time through chemical processes. Luminescence dating, with its focus on trapped electrons and their release, brought to light both the ingenuity and the challenges of dating sediments and artifacts. The discussions on other methods further emphasized the diversity and specificity of dating techniques, each with its unique applicability and set of limitations.

This chapter has underscored a crucial aspect of scientific inquiry – the necessity for skepticism and critical analysis. The debates and controversies surrounding these dating methods are not merely academic exercises but are fundamental to the dynamic nature of scientific progress. They remind us that science is not a static field of known facts but a continuous quest for understanding, often marked by revisions and refinements.

In the area of evolutionary theory and archaeology, the implications of dating inaccuracies can be profound. Misdated artifacts or geological layers can lead to misinterpretations of human history or the Earth's past, affecting our understanding of everything from human migration patterns to climate changes over millennia. Thus, it becomes imperative to approach these dating methods not as infallible solutions but as tools that, while powerful, require cautious and knowledgeable interpretation.

As we move forward in our scientific endeavors, the lessons from this chapter highlight the need for a multidisciplinary approach, one that combines various dating methods with a thorough understanding of the physical and chemical processes at play. It also advocates for an open-minded yet critical approach to scientific findings, where questioning and re-evaluation are seen as integral to the advancement of knowledge.

The attempt to measure time, to place our history within the framework of years, centuries, and millennia, is as much about understanding the limitations of our tools as it is about the discoveries they enable. As we continue to explore the mysteries of the universe and our place within it, let us remember that skepticism is not just a philosophical stance but a practical necessity in the pursuit of truth.

CHAPTER 3

THE FOSSIL RECORD: WHAT IT TELLS US AND WHAT IT DOESN'T

In our attempt to understand the history of life on Earth, few tools have been as influential as the fossil record. It stands as a testament to the long and complex history of life, offering us a window into an ancient world that would otherwise remain shrouded in mystery. Yet, the story it tells is not without its gaps and ambiguities.

At its core, the fossil record is an accumulation of the remains or traces of organisms from the past, embedded in rock layers over millions of years. These fossils, ranging from the imprints of the most diminutive organisms to the bones of colossal dinosaurs, have been pivotal in piecing together the puzzle of evolutionary history. They provide crucial evidence for the changes in life forms over time, supporting the theory of evolution by natural selection as proposed by Charles Darwin and others.

However, as we delve deeper into this chapter, we adopt a lens of skepticism to examine the fossil record. This approach is not to undermine its significance but to understand its limitations and the extent to which it can reliably inform us about the past. The process of fossilization is selective and rare, meaning that the fossil record is inherently incomplete. Not all organisms have an equal chance of being preserved as fossils, leading to gaps that can be difficult to fill.

Moreover, the interpretation of fossils is often influenced by the prevailing scientific theories and biases of the time. Controversies and debates have arisen over certain fossil finds, challenging our understanding of evolutionary timelines and relationships. Critics of evolutionary theory have pointed to these gaps and inconsistencies as weaknesses in the narrative of life's history.

This chapter will explore the formation of fossils, the different types that exist, and what they reveal about past life. We will critically examine the notable fossil finds that have shaped our understanding, the gaps and inconsistencies in the record, and the controversial discoveries that continue to spark debate. By presenting skeptical viewpoints and critiques from scientists, we aim to provide a balanced view of the fossil record, acknowledging its contributions to science while also recognizing its limitations and the questions it leaves unanswered.

As we navigate through this chapter, we invite readers to engage with the fossil record not just as a collection of ancient remains, but as a complex and sometimes contentious chapter in the story of life on Earth, one that is still being written and reinterpreted.

FORMATION OF FOSSILS

The process of fossilization is a remarkable and intricate one, a confluence of time, geology, and biological happenstance that turns once-living organisms into the stone echoes of the past. Understanding this process is crucial for comprehending what the fossil record can - and cannot - tell us about the history of life on Earth.

The Rarity of Fossilization:

Fossilization is an exceptionally rare event. The vast majority of organisms that have ever lived on Earth have left no trace in the fossil record. This is due to the precise and often fortuitous set of conditions required for an organism to become fossilized. The rarity of these conditions immediately introduces a significant bias into the fossil record - it is far from a complete account of life's history.

Initial Burial:

The first step in the journey towards fossilization is burial. After an organism dies, it must be quickly buried in sediment to protect it from scavengers, decay, and environmental elements. This burial can occur in various environments, such as riverbeds, lakes, oceans, or volcanic ash. The speed and manner of burial significantly affect the chances of fossilization.

Mineralization:

The most common fossilization process is mineralization. Over time, the organic materials in the organism, like bones or wood, are replaced with minerals from the surrounding environment. This process, known as permineralization, occurs as groundwater rich in minerals like silica or calcium carbonate permeates the buried remains. The minerals precipitate out of the water and fill the cellular spaces and tiny pores of the original organism, creating a stone replica.

Preservation Conditions:

The likelihood of fossilization is influenced by several factors, including the organism's environment, its physical and chemical makeup, and the conditions after death. Hard parts like bones and shells are more likely to fossilize than soft tissues. Acidic environments, which dissolve bone and shell, are less conducive to fossilization. Anoxic environments, where oxygen is scarce, are more favorable because they slow down decay and scavenging.

Taphonomy:

The study of how organisms decay and become fossilized is known as taphonomy. Taphonomic research helps scientists understand not just the process of fossilization but also the biases and gaps in the fossil record. It reveals how factors like transport after death, scavenging, and environmental changes can affect the final fossil.

The formation of fossils is a process heavily influenced by environmental factors and chance. This understanding leads to a recognition of the inherent incompleteness and selectivity of the fossil record. While fossils provide invaluable insights into life's past, the story they tell is fragmented and subject to the whims of geological and biological processes.

Types of Fossilization:

Fossilization can occur in several distinct forms, each capturing a different aspect of an organism's existence and contributing uniquely to the fossil record. Understanding these various types helps in comprehending the diversity and limitations of the information that fossils can provide.

Permineralization: One of the most common forms of fossilization, involves the gradual infiltration of mineral-rich water into the porous parts of an organism's remains, such as bones or wood. The minerals precipitate from the water, filling the pores and crystallizing to form a stone-like replica of the original biological material. Petrification is a similar process where the organic material is completely replaced by minerals, often retaining detailed structures of the organism.

Molds and Casts: Molds and casts are indirect forms of fossilization. A mold is created when an organism is buried in sediment and then decays, leaving behind an impression of itself. If this mold is later filled with other sediments or minerals, a cast is formed. This process can preserve detailed external features of the organism, but not its internal structures.

Carbonization (Carbon Films): Carbonization occurs when an organism is subjected to pressure and heat over time, causing all the volatile elements except carbon to dissipate. What remains is a thin layer of carbon residue, often preserving fine details of soft-bodied organisms like leaves, fish, and insects. This method is particularly significant for studying the morphology of ancient plants and delicate organisms.

Trace Fossils (Ichnofossils): Trace fossils, or ichnofossils, are records of an organism's activity rather than its physical remains. They include footprints, burrows, feeding marks, and feces (coprolites). Trace fossils are crucial for understanding the behavior, movement, and interaction of organisms with their environment, providing insights that body fossils alone cannot.

Resin Fossils (Amber): Occasionally, organisms or parts of organisms are trapped in tree resin, which later hardens into amber. This type of fossilization is excellent for preserving fine details of small organisms like insects. The clarity with which these organisms are preserved offers a unique window into ancient ecosystems.

Bioimmuration: Bioimmuration occurs when an organism is overgrown by another organism, typically a sessile marine organism like a coral. The overgrowing organism leaves an impression of the buried organism, providing information about its presence even if the original organism has decayed.

Each type of fossilization captures a different aspect of ancient life, from the organism's physical form to its interactions with the environment. This diversity is both a strength and a limitation of the fossil record. It allows scientists to reconstruct various facets of past ecosystems but also means that certain types of organisms and activities are more likely to be preserved than others.

WHAT PROBLEMS THE FOSSIL RECORD REVEALS

The fossil record, while often cited as a bedrock of evolutionary biology, also presents a series of problems and challenges that raise critical questions about our understanding of life's history. This section explores the inherent issues and limitations within the fossil record that complicate its interpretation in the context of evolutionary theory.

Incomplete and Fragmentary Nature:

>   » One of the most significant problems with the fossil record is its incomplete and fragmentary nature. The process of fossilization is rare and selective, favoring certain organisms, environments, and conditions over others. This selectivity leads to substantial gaps in the record, particularly in the early stages of life and across transitions between major groups. Such gaps pose a challenge in reconstructing a continuous and comprehensive narrative of evolutionary history.

Misinterpretation and Revision:

>   » The history of paleontology is marked by instances of misinterpretation, where fossil findings initially thought to support certain theories were later reassessed and

understood differently. This shifting landscape of interpretation, exemplified by cases like the Piltdown Man hoax, underscores the fallibility inherent in scientific study and the influence of biases and preconceived notions.

Debate over Transitional Forms:

» Transitional fossils, often heralded as key evidence for evolutionary transitions, are themselves a source of debate. Critics argue that the scarcity of clear transitional forms in the fossil record points to problems in the theory of gradual evolutionary change. Proponents contend that such transitions are rarely captured due to the sporadic nature of fossilization, but critics see this as a convenient explanation for missing data.

Patterns of Stasis and Rapid Change:

» The fossil record shows long periods of stasis (little or no evolutionary change) punctuated by relatively rapid changes, a pattern encapsulated in the theory of punctuated equilibrium. This observation challenges the traditional view of gradual, steady evolutionary change and opens up debates about the mechanisms driving evolutionary processes.

Ambiguity in Interpretation:

» Fossils, being remnants of past life, are open to interpretation. Determining the ecology, behavior, or even the exact taxonomy of fossilized organisms can be fraught with uncertainty. This ambiguity extends to major events, like mass extinctions, where the exact causes and consequences are still subjects of active research and debate.

The fossil record, though invaluable as a historical document of life on Earth, reveals a series of problems that challenge straightforward interpretations. Its gaps, the rarity of fossilization, the potential for misinterpretation, and the complex patterns it reveals demand a cautious and critical approach. These challenges remind us that the fossil record is not a complete storybook of life's history but a fragmented collection of chapters, each subject to revision and reinterpretation.

Examples of misinterpreted fossils

There have been several notable examples in the history of paleontology where fossil finds were initially misinterpreted, or where certain discoveries have posed challenges to prevailing theories of evolution. Here are a few examples:

Piltdown Man:

> » Perhaps the most infamous case of misinterpretation in the history of paleontology is the Piltdown Man, discovered in England in 1912. Initially, it was hailed as a crucial missing link in human evolution, combining a human-like skull with an ape-like jaw. However, in 1953, it was revealed to be a hoax, combining a medieval human skull with the lower jaw of an orangutan. This case stands as a stark reminder of the potential for error and fraud in paleontological interpretation.

Archaeopteryx:

» Archaeopteryx, often cited as a classic transitional fossil between dinosaurs and birds, has been subject to ongoing debate and reinterpretation. While it clearly shows a mix of avian and reptilian features, some researchers have questioned its role in the direct lineage of modern birds. The discovery of other feathered dinosaurs has further complicated the picture, leading to debates about the exact evolutionary relationship between birds and dinosaurs.

Neanderthals:

» Early interpretations of Neanderthals portrayed them as brutish and primitive, significantly different from modern humans. However, ongoing research, including genetic studies, has radically changed our understanding of Neanderthals. They are now seen as challenging earlier notions of a linear, progressive evolution of modern humans.

Coelacanth:

» The discovery of the living coelacanth in 1938 was a significant surprise to the scientific community. Previously, coelacanths were known only from fossils and thought to have gone extinct around 66 million years ago. The existence of living specimens challenged assumptions about extinction events and the completeness of the fossil record.

Cambrian Explosion:

> » The Cambrian Explosion refers to a period approximately 541 million years ago when a sudden burst of life forms appeared in the fossil record. This event has been challenging to reconcile with the theory of gradual evolution, as it seems to suggest a rapid emergence of diverse, complex organisms. While various hypotheses have been proposed, including changes in environmental conditions and genetic developments, the Cambrian Explosion remains a subject of intense study and debate.

These examples illustrate how fossil discoveries can both illuminate and challenge our understanding of evolutionary history. They underscore the importance of continual reassessment and critical analysis in the interpretation of the fossil record.

Gaps and Inconsistencies in the Fossil Record

The fossil record, while a rich tapestry of ancient life, is punctuated by notable gaps and inconsistencies. These missing pieces, often referred to as "missing links," have been at the center of significant debate and scrutiny, particularly in the context of evolutionary theory.

The Nature of Missing Links:

The concept of 'missing links' in the fossil record is a pivotal and often contentious aspect of paleontological research, particularly in its relationship with evolutionary theory. These missing links, or transitional fossils, are critical pieces in the puzzle of life's evolutionary history, yet their elusive nature presents a complex challenge to our understanding.

Missing links are theoretically those fossils that exhibit traits of both an ancestral group and its derived descendant, effectively bridging gaps in the morphological record. They are the physical manifestations of evolutionary theory, the tangible evidence of change over time. In an ideal scenario, these fossils would neatly connect the dots between major groups of organisms, illustrating a clear line of descent.

Evolutionary theory, particularly as it was initially conceptualized, suggests a gradual, linear progression of change over time. Under this framework, missing links are expected to be abundant and clear-cut. However, the reality presented by the fossil record is markedly different. Transitional fossils are not always clearly demarcated, and the evolutionary lineage they suggest is often far from linear. This discrepancy between theoretical expectation and actual evidence fuels much of the debate surrounding missing links.

Identifying and classifying transitional fossils is fraught with challenges. Morphological characteristics may not be clearly indicative of evolutionary relationships, and the interpretation of these traits can be subjective, influenced by the prevailing scientific theories and biases. A fossil thought to be a transitional form in one era of scientific thought may be reinterpreted differently as new theories and information emerge.

Throughout the history of paleontology, there have been numerous fossils hailed as missing links, only to become subjects of controversy and reevaluation. For example, the discovery of Archaeopteryx, with its blend of avian and reptilian features, was initially seen as a clear transitional form between dinosaurs and birds. Over time, however, the discovery of other feathered dinosaurs has complicated its status, leading to debates about the exact nature of bird evolution.

The sporadic and often ambiguous nature of missing links in the fossil record challenges some traditional perspectives on evolution. It raises questions about the pace and process of evolutionary change, suggesting a more complex and branching evolutionary history than a simple, direct lineage. This complexity does not necessarily refute evolutionary theory but highlights the nuanced and multifaceted nature of biological change over time.

The nature of missing links in the fossil record is a nuanced and complex topic. The rarity and ambiguity of these transitional fossils challenge simplistic views of evolution and underscore the dynamic, non-linear nature of life's history. As such, they are a source of ongoing research, debate, and reinterpretation within the scientific community, continually reshaping our understanding of evolutionary processes.

Implications for Evolutionary Theory

The gaps and so-called missing links in the fossil record have significant implications for evolutionary theory, especially regarding the nature and pace of evolutionary change. These implications are a focal point for critics who challenge the traditional narrative of evolution as proposed by Darwin.

Central to Darwin's theory of evolution is the concept of gradualism – the idea that evolutionary change occurs slowly and incrementally over vast periods of time. According to this view, the transition from one species to another or the emergence of new anatomical features should be a gradual process, leaving behind a trail of intermediate forms. Critics of evolutionary theory seize upon the gaps in the fossil record, arguing that the absence of a continuous chain of transitional fossils undermines the concept of gradual, steady evolutionary change. They contend that if gradualism were the rule, the fossil record should be abundant with clear examples of these intermediate forms.

The rarity of fossils that clearly represent transitional forms between major evolutionary groups is a point of contention. Critics argue that this rarity is inconsistent with the predictions of gradualistic evolution. They suggest that the sporadic appearance of major groups in the fossil record, without a clear sequence of transitional forms, challenges the notion of a smooth, linear evolutionary progression.

Proponents of evolutionary theory often argue that the incompleteness of the fossil record is a primary reason for the apparent gaps. They point out that fossilization is a rare event, influenced by a multitude of ecological and geological factors, and thus the fossil record is not expected to capture every step of evolutionary change. This viewpoint posits that the gaps in the record are an artifact of the imperfect nature of fossil preservation rather than evidence against evolutionary theory.

The debate over these gaps and missing links has had a profound impact on evolutionary research. It has driven continuous efforts to find new fossils, develop better dating methods, and refine evolutionary models. The discourse around these gaps underscores the dynamic nature of scientific inquiry, where theories are continually tested, challenged, and revised in light of new evidence.

The gaps and missing links in the fossil record, and their implications for evolutionary theory, are subjects of ongoing scientific debate and research. They highlight the complex and often non-linear nature of the evolutionary process, challenging researchers to continually refine their understanding of life's history on Earth.

Controversial Finds

Certain discoveries stand out not just for what they reveal, but also for the debates and controversies they ignite. These controversial finds often serve as flashpoints in the ongoing discourse of evolutionary biology, challenging established theories and prompting new questions about the history of life on Earth.

The nature of a controversial fossil find is multifaceted. Often, these fossils are unexpected or anomalous, defying the established predictions and assumptions of evolutionary theory. They may represent an organism or a trait that appears out of place in the understood timeline of life, or they might contradict prevailing ideas about the phylogenetic relationships between different species. In other instances, the controversy stems from the interpretation of the fossil itself – its age, its classification, or the inferences drawn about its biology and ecology.

The discovery of such fossils frequently sets off a ripple of scientific debate. These debates are not merely academic; they are fundamental to the progress of science, testing and refining our understanding of evolutionary processes. They represent the dynamic interplay between evidence and theory, where new findings can either reinforce or reshape the existing body of knowledge.

For instance, the discovery of a fossil with characteristics that appear to bridge two distinct groups can challenge the understanding of how those groups evolved and diverged. Alternatively, a fossil that seems to belong to a group thought to be extinct can raise questions about the completeness and accuracy of the fossil record.

Controversial fossil finds thus play a critical role in the story of evolutionary biology. They are catalysts for scientific discussion and inquiry, pushing researchers to re-examine and reassess long-held views. In exploring these finds, we delve into a realm where science is as much about grappling with uncertainties and unknowns as it is about accumulating knowledge and facts. It is this interplay of discovery and debate that drives the evolution of our understanding of life's past.

Detailed Examples of Controversial Fossils

Each of the following examples represents a key moment in paleontological discovery, where the unearthed fossils challenged conventional thinking and ignited scientific debate:

Homo Naledi:

The discovery of Homo naledi in the Rising Star Cave system in South Africa marked a significant and controversial moment in the study of human evolution. Unearthed in a complex, hard-to-reach cave chamber, these fossils presented a perplexing mix of both primitive and modern traits that stirred intense debate in the scientific community.

Homo naledi's anatomy was unique. The creature had a small brain akin to early human ancestors, yet its hands and feet bore striking resemblances to modern humans. This combination of features was unlike anything previously seen in the human fossil record. Its curved fingers suggested climbing ability, while its feet implied it was adept at walking upright.

Initially, there was no clear indication of the age of the Homo naledi fossils. This uncertainty fueled debate about where, or even if, these beings fit into the human evolutionary timeline. When dating finally suggested that Homo naledi lived between 236,000 and 335,000 years ago, it raised profound questions. This time period meant that Homo naledi could have coexisted with anatomically modern humans, challenging the notion of a linear progression from primitive to modern human forms.

The discovery was met with skepticism by some in the scientific community. Critics questioned whether Homo naledi represented a new species or was merely a variant of a known species, possibly exhibiting pathological characteristics. Others pointed to the method of its discovery and the initial lack of dating as complicating factors in understanding its significance. The debate was not just about Homo naledi's physical characteristics but also about its behavior, with some researchers proposing that the location and manner of the fossil deposit suggested ritualistic burial, a behavior long thought to be unique to Homo sapiens.

Despite the controversies, the discovery of Homo naledi has been pivotal in highlighting the diversity and complexity of the human lineage. It challenges the traditional, more straightforward view of human evolution, suggesting a more intricate web of ancestral and descendant species. Homo naledi's place in the human story continues to be a subject of active research and debate, embodying the evolving nature of our understanding of human origins.

The case of Homo naledi underscores the dynamic and sometimes contentious nature of paleoanthropological research.

Feathered Dinosaurs and the Bird-Dinosaur Link

The discovery of feathered dinosaurs in the late 20th and early 21st centuries fundamentally altered our understanding of the relationship between birds and dinosaurs. These finds, primarily in China, have provided compelling evidence supporting the theory that birds evolved from theropod dinosaurs. However, the interpretation of these feathered specimens and their place in the evolutionary tree has been a subject of intense debate.

The first well-preserved specimens of feathered dinosaurs, such as Sinosauropteryx and later, the more bird-like Archaeopteryx, revealed a complex mosaic of features. These dinosaurs had distinct dinosaurian traits, yet bore feathers — a characteristic long thought to be exclusive to birds. The presence of feathers on these theropods suggested a direct evolutionary link between birds and dinosaurs.

The idea that birds could be the modern descendants of dinosaurs was initially met with skepticism. This was a radical shift from the traditional view of birds and reptiles as distinct, unrelated groups. Critics questioned the interpretation of the feathers, some suggesting they were not true feathers but rather simple filamentous structures or a result of fossilization processes.

As more feathered dinosaur specimens were unearthed, each with varying features, debates intensified about the evolution of flight, bird-like behaviors in dinosaurs, and the exact point in the dinosaur lineage where birds diverged. The diversity of feathered dinosaurs raised questions about whether feathers initially evolved for flight or other functions like insulation or display.

The discovery of feathered dinosaurs has implications for the phylogenetic tree — how different groups of organisms are related through evolutionary history. Some paleontologists argue that these findings support a radical reorganization of the dinosaur-bird segment of the tree, while others caution against hasty revisions without more conclusive evidence.

Ongoing research continues to explore the complexities of this evolutionary transition, examining new fossil finds, and employing advanced technologies like CT scans and molecular analysis to gain deeper insights into the bird-dinosaur link.

The case of feathered dinosaurs and their role in understanding the bird-dinosaur link is a striking example of how new discoveries can challenge established scientific narratives. It reflects the dynamic nature of paleontological research, where each new find can prompt a reevaluation of long-standing theories and ignite fresh debates in the quest to understand life's evolutionary journey.

The 'Hobbit' - Homo Floresiensis

The discovery of Homo floresiensis on the Indonesian island of Flores in 2003 added a new, intriguing chapter to the story of human evolution. The diminutive stature and unusual anatomical features of these individuals quickly earned them the nickname 'Hobbits' and sparked a flurry of scientific debate and skepticism about their place in the human evolutionary tree.

The most striking feature of Homo floresiensis was its small size, with adults standing about 3.5 feet tall and having a brain size comparable to chimpanzees. Initially discovered in the Liang Bua cave, these remains were initially thought to be around 18,000 years old, placing them in relatively recent history, potentially overlapping with modern humans.

The classification of Homo floresiensis has been a matter of intense debate. Some scientists proposed that they represented a new species in the human lineage, possibly a descendant of Homo erectus but with unique adaptations due to island dwarfism. Others, however, were skeptical, suggesting that the individuals were modern humans (Homo sapiens) suffering from pathological conditions like microcephaly or endemic cretinism, which could account for their small stature and brain size.

If Homo floresiensis were indeed a separate species, their existence challenged conventional theories about human evolution, particularly the idea that larger brain size was a consistent trend in hominid development. The notion of a small-brained human species surviving until relatively recent times raised questions.

Ongoing research and more recent discoveries have shed additional light on these debates. Studies of the wrist bones and other anatomical features have supported the distinct species hypothesis. Additionally, revised dating methods placed the Hobbit remains at around 50,000 to 60,000 years old, with evidence suggesting they could have lived as recently as 50,000 years ago, further complicating the timeline of human evolution.

In summary, Homo floresiensis remains one of the most enigmatic and debated figures in human evolutionary research.

Ardipithecus Ramidus:

Ardipithecus ramidus, a hominid fossil dating back 4.4 million years, challenged ideas about early human evolution, particularly the nature of our last common ancestor with chimpanzees. Its mix of bipedal and arboreal traits contradicted the then-prevailing notion that early hominids were very chimp-like. The interpretation of Ardi's features and its place in the human lineage sparked debates over the mechanisms of human evolution.

Each of these controversial finds has added a layer of complexity to our understanding of evolution. They challenge us to reconsider what we think we know about the path of life on Earth, highlighting the dynamic and ever-evolving nature of scientific discovery.

CHAPTER 4

DECODING EVOLUTION – A CRITICAL EXAMINATION OF GENETICS AND MOLECULAR BIOLOGY

As we embark on the journey through Chapter 4 of "Rethinking Science: Questioning Evolution and Theories of the Universe," we delve into the intricate world of genetics and molecular biology, the bedrock upon which much of modern evolutionary theory is built. This chapter aims to unravel the complex tapestry of DNA, genes, and the molecular mechanisms that underpin life, while casting a critical and skeptical eye on how these elements fit into the broader narrative of evolution.

The story of life, as told through the lens of genetics, is one of intricate codes, molecular interactions, and the transfer of biological information across generations. DNA, the molecule at the heart of this narrative, has often been portrayed as the master key to understanding life's diversity and the process of evolution itself. Yet, beneath the surface of this seemingly straightforward story lies a realm of complexity and nuance that prompts us to question and reassess.

In this chapter, we will explore the foundational concepts of genetics – the structure and function of DNA, the mechanisms of genetic inheritance, and the role of mutations in driving evolutionary change. We will confront the challenges to the conventional DNA-centric view of evolution, examining the interplay between genes, environment, and developmental processes. Our critical examination will extend to the realm of molecular biology, where we scrutinize the molecular basis of evolution and the debates surrounding it.

This chapter is not just an exploration of genetic and molecular biology concepts; it is an invitation to engage with these topics critically. We will present skeptical viewpoints that challenge the sufficiency of genetics in explaining the complexity of life and evolutionary processes. Through case studies, examples, and critiques from scientists, we aim to provide a balanced perspective that recognizes the contributions of genetics to our understanding of evolution, while also highlighting the limitations and unanswered questions in this field.

As we decode the mysteries of genetics and molecular biology, we are reminded that science is a continuously evolving pursuit of knowledge. What we understand today may be redefined tomorrow, and it is this dynamic nature of scientific inquiry that lies at the heart of our exploration.

SECTION 1

BASICS OF DNA AND GENETICS

At the heart of every living organism, from the simplest bacteria to the most complex human, lies the molecule of life: deoxyribonucleic acid, commonly known as DNA. This remarkable molecule holds the instructions necessary for the development, functioning, and reproduction of all living beings. Understanding the structure, function, and role of DNA is crucial for delving into the molecular basis of genetics and evolution.

DNA is a long, double-stranded molecule composed of nucleotides. Each nucleotide consists of a sugar (deoxyribose), a phosphate group, and a nitrogenous base. The sequence of these bases (adenine, thymine, guanine, and cytosine) along the DNA strand encodes the genetic information of an organism.

The structure of DNA is famously characterized by the double helix model, where two strands twist around each other, held together by hydrogen bonds between complementary base pairs (adenine with thymine, and guanine with cytosine).

A gene is a segment of DNA that contains the instructions for making a specific protein or set of proteins. Proteins, in turn, are responsible for most functions in a living organism. The human genome, for instance, contains about 20,000-25,000 genes, although the number of genes varies widely between species.

These genes are organized into structures called chromosomes. Humans, for example, have 23 pairs of chromosomes in each cell, with one set inherited from each parent. Chromosomes ensure that DNA is accurately replicated and distributed during cell division.

The process of gene expression involves the conversion of information from a gene into a functional product, typically a protein. This process involves two main steps: transcription, where a segment of DNA is copied into messenger RNA (mRNA), and translation, where the mRNA is used as a template to assemble amino acids into a protein.

Gene expression is not a static process; it is finely regulated by the organism in response to various internal and external signals. This regulation is critical for processes like development, response to the environment, and maintaining homeostasis.

The slight differences in the DNA sequences among individuals within a species are the basis of genetic variability. This variability is considered crucial for evolution, as it provides the raw material upon which natural selection acts. Mutations, which are changes in the DNA sequence, can introduce new genetic variations and potentially new traits into a population.

DNA and its organization into genes and chromosomes form the foundation of genetic inheritance and variation, playing a pivotal role in the processes of life and evolution.

Mechanisms of Genetic Inheritance

The passage of genetic information from one generation to the next is a fundamental process that underpins the continuity and diversity of life. This intricate mechanism, involving DNA replication, transcription, translation, and the role of RNA, ensures that genetic information is not only preserved but also allows for variation, which is a key driver of evolution.

DNA replication is the process by which a cell duplicates its DNA before cell division, ensuring that each new cell receives a complete set of genetic instructions. This process begins with the unwinding of the double helix, followed by the synthesis of new DNA strands complementary to the original strands. Enzymes like DNA polymerase play a crucial role in this process, adding nucleotides one by one to build the new strands. Replication is remarkably accurate, but errors can occur, leading to mutations that may have evolutionary consequences.

Transcription is the first step in gene expression, where the information in a gene's DNA is transferred to a messenger RNA (mRNA) molecule. This process involves the enzyme RNA polymerase, which binds to a specific region on the DNA and synthesizes an mRNA strand by linking RNA nucleotides that are complementary to the DNA template. The resulting mRNA serves as a temporary copy of the genetic information, carrying the code needed to build proteins.

Translation is the process by which the genetic code carried by mRNA is used to build proteins. This occurs in the cell's ribosomes, where transfer RNA (tRNA) molecules bring specific amino acids in sequence according to the mRNA code. Each set of three nucleotides on the mRNA, called a codon, corresponds to a specific amino acid or a signal to stop protein synthesis. The sequence of amino acids thus assembled forms a polypeptide chain, which folds into a functional protein.

While RNA is well-known for its role in transcription and translation, it also performs various other critical functions in cells. For instance, ribosomal RNA (rRNA) and transfer RNA (tRNA) are key components of the protein synthesis machinery. Moreover, recent discoveries have revealed the importance of various types of non-coding RNAs in regulating gene expression and maintaining genomic integrity.

Genetic variation, essential for evolution, arises through processes such as mutation and genetic recombination. Mutations are changes in the DNA sequence that can be caused by errors during DNA replication or external factors like radiation. While most mutations are neutral or harmful, occasionally, a mutation can confer a beneficial trait that can be selected for in a population.

Genetic recombination, which occurs during sexual reproduction, involves the exchange of genetic material between chromosomes, leading to offspring with a unique combination of genes from their parents. This recombination contributes to genetic diversity within a population, providing a vast pool of traits upon which natural selection can act.

The mechanisms of genetic inheritance are central to the continuity and evolution of life. Through processes like DNA replication, transcription, translation, and the action of various RNA molecules, genetic information is accurately transmitted and expressed in living organisms. Moreover, the generation of genetic variation through mutation and recombination is fundamental to the process of evolution, providing the raw material for natural selection and adaptation.

Skeptical Analysis of Genetic Determinism

Defining Genetic Determinism

In exploring the concept of genetic determinism, we venture into a domain where genetics is often perceived as the master architect of life. Genetic determinism is the theory that an organism's genetic makeup, encoded in the DNA, is the primary determinant of all physical and behavioral traits. According to this view, genes are the blueprint from which all characteristics, from the color of one's eyes to predispositions for certain diseases, and even behavioral traits, are derived. In essence, it posits that the complex tapestry of life can be unraveled by decoding the sequences of nucleotides in DNA.

This concept is deeply rooted in the molecular perspective of biology, which gained prominence with the discovery of the structure of DNA and the understanding of its role in heredity. The appeal of genetic determinism lies in its simplicity; it offers a clear and direct mechanism by which traits are passed from generation to generation and by which evolutionary changes can ostensibly occur. It suggests a direct causality where specific genes or combinations of genes can be linked directly to specific traits or behaviors.

However, this straightforward relationship between genes and traits has been a subject of intense debate and skepticism. Critics argue that genetic determinism oversimplifies the intricate workings of biological systems. Biology, they contend, is not a simple readout of a genetic script but a complex interplay of

117

genetic, environmental, and developmental factors. They point out that while genes undoubtedly play a crucial role in development and heredity, they do not act in isolation. The environment in which an organism develops and lives can significantly influence how genes are expressed and function.

Moreover, the notion of genetic determinism has been challenged by emerging fields like epigenetics, which studies heritable changes in gene expression that do not involve alterations to the DNA sequence. Epigenetic mechanisms, such as DNA methylation and histone modification, demonstrate that gene expression can be modified by environmental factors, adding layers of complexity to the relationship between genes and phenotypes.

In the context of evolution, genetic determinism has been scrutinized for its implications in understanding the mechanisms of evolutionary change. If genes alone determined traits, evolutionary changes might be expected to follow a more predictable and linear path. However, the fossil record and observations of living organisms reveal a more intricate and non-linear process of evolution, suggesting that other factors play significant roles in shaping the course of evolutionary history.

While the concept of genetic determinism provides a foundational understanding of the role of genes in biology, it is increasingly viewed as an incomplete explanation for the diversity and complexity of life. This section of the chapter sets the stage for a deeper examination of the critiques and limitations of genetic determinism, encouraging readers to consider a more holistic view of genetics, environment, and development in the tapestry of life.

Historical Context and Evolutionary Implications of Genetic Determinism

The concept of genetic determinism, while gaining significant traction in the 20th century, has historical roots that intertwine with the early understanding of heredity and evolution. This perspective emerged prominently with the rediscovery of Mendel's work on inheritance in the late 19th century and was further cemented with the rise of molecular biology and the identification of DNA as the carrier of genetic information in the mid-20th century. The allure of genetic determinism lies in its apparent clarity and precision – the notion that the complex processes of life can be traced back to specific sequences within the DNA.

From Mendel to the Modern Synthesis:

» Gregor Mendel's experiments with pea plants laid the foundation for modern genetics, demonstrating that traits are inherited in predictable patterns. However, Mendel's work initially focused on discrete, easily categorizable traits, inadvertently paving the way for a simplified view of genetics.

» The Modern Synthesis of the 1930s and 1940s, which integrated Darwinian evolution with Mendelian genetics, further reinforced the role of genetics in evolution. This synthesis highlighted natural selection acting on genetic variations as the primary driver of evolutionary change. Yet, this framework, while groundbreaking, often leaned towards a gene-centric view of evolution, sometimes overlooking the complexities of organism-environment interactions.

Challenges to a Linear Perspective:

The linear perspective of genetic determinism has been increasingly challenged in light of discoveries that reveal the multi-layered nature of gene-environment interactions. Critics argue that the Modern Synthesis, while powerful, oversimplified the evolutionary process by not fully accounting for these interactions and the role of developmental processes in evolution.

The discovery of phenomena such as epigenetic inheritance, where environmental factors can cause changes in gene expression that are heritable, has further complicated the narrative. These findings suggest that the environment can directly influence the genetic legacy of organisms, a concept not readily accommodated within the strict boundaries of genetic determinism.

The implications of moving beyond a strictly gene-centric view are profound for evolutionary theory. It calls for a more nuanced understanding of evolution, one that considers not just the changes in gene frequencies over time, but also how genes interact with each other and with their environments to produce the diversity of life.

This perspective acknowledges that evolutionary change is not merely a product of random mutations filtered by natural selection, but a complex process influenced by a myriad of factors including genetic networks, environmental conditions, and developmental pathways.

While genetic determinism provided a critical framework for understanding the role of genetics in evolution, its historical context and recent scientific advancements invite a reevaluation. It is now increasingly recognized that a comprehensive understanding of evolution requires integrating genetics with ecological, environmental, and developmental factors. This section of the book underscores the importance of considering these broader interactions and challenges us to rethink the narrative of evolution.

Challenges to Genetic Determinism

As we delve deeper into the nuances of genetic determinism, it becomes evident that this theory faces significant challenges and critiques. Central to these is the argument that genetic determinism grossly oversimplifies the intricate interplay between genes, the environment, and developmental processes, which collectively shape an organism's traits and behaviors. This section examines these critiques, highlighting how they question the reductionist view that genes alone dictate biological outcomes.

One of the fundamental criticisms of genetic determinism is its failure to adequately account for the complex interactions between genes and environmental factors. The expression of genes is not an isolated process but is heavily influenced by the environment in which an organism lives. Factors such as nutrition, stress, exposure to toxins, and social interactions can profoundly impact how genes are expressed. This dynamic interaction suggests that the same genetic makeup can lead to different outcomes under different environmental conditions, challenging the notion of a direct, one-to-one relationship between genes and traits.

Epigenetics has emerged as a key field challenging genetic determinism. It involves changes in gene expression that do not alter the DNA sequence but are still heritable. These changes are often triggered by environmental factors and can have lasting effects on gene function. Epigenetic mechanisms, like DNA methylation and histone modification, demonstrate that an organism's environment can leave molecular marks on genes that affect their activity. This discovery has profound implications, suggesting that the environment can shape heredity in ways that genetic determinism does not predict.

Another critique focuses on the complexity of genetic interactions within an organism. Genes do not work in isolation; they function as part of complex networks where multiple genes and their products interact with each other. This complexity means that pinpointing a single gene for a specific trait is often an oversimplification. It also implies that the effects of a genetic mutation can depend on the context of the entire genome, rather than just the change in a single gene.

Developmental plasticity, the ability of an organism to change its developmental course in response to environmental conditions, further challenges genetic determinism. This plasticity demonstrates that the developmental outcome is not rigidly programmed by genes but can be significantly influenced by external factors during an organism's growth and development.

The critiques of genetic determinism have significant implications for evolution theories. They suggest that evolutionary change is not just a matter of genetic mutations being selected over generations but also involves a more complex interplay of genetics, environment, and developmental processes. This perspective encourages a broader view of evolution.

In conclusion, the challenges to genetic determinism highlight the need for a more comprehensive approach to understanding the relationship between genes, environment, and development. These critiques urge us to move beyond a simplistic gene-centric view of biology and embrace a more integrated perspective that acknowledges the complexity of life.

Case Studies and Counterexamples

In the ongoing discourse about genetic determinism and its role in defining traits and behaviors, several compelling case studies and counterexamples emerge that challenge the notion of genes as the sole architects of biological destiny. These instances underscore the complex interplay of genetics, environment, and individual development.

Monozygotic Twins: A Study in Genetic Similarity and Variability

One of the most striking illustrations of the limitations of genetic determinism is observed in monozygotic, or identical, twins. These individuals, who develop from a single fertilized egg, share nearly identical genetic makeup. However, despite these genetic similarities, numerous studies have shown that monozygotic twins can exhibit significant differences in traits, health conditions, and even behaviors.

For instance, while one twin may develop a genetic disorder like schizophrenia or type 1 diabetes, the other twin does not always manifest the same condition, despite their identical DNA. This divergence is particularly intriguing and poses a significant challenge to the idea that genes alone dictate our biological and behavioral destinies. It suggests that while genetics play a crucial role in laying down the foundations of our traits, they do not unilaterally determine our health, personality, or behavior.

The twin studies have prompted researchers to explore the role of environmental factors and unique experiences that each twin encounters. These factors include variations in the womb environment, differences in life experiences, diets, exposure to pathogens, and even varying levels of stress. The differences in the outcomes among monozygotic twins serve as a powerful testament to the influence of environmental factors in shaping our traits and health conditions.

Environmental Influences: Epigenetic Modifications

Environmental impact on gene expression is further evidenced by the field of epigenetics. Epigenetic changes are modifications that alter gene activity without changing the DNA sequence. These changes can be triggered by environmental factors and, remarkably, can be passed down to subsequent generations.

A notable example is the Dutch Hunger Winter of 1944-1945, a tragic event that has provided valuable insights into the field of epigenetics. Studies of individuals who were in utero during this famine revealed that they experienced higher rates of health issues like obesity, diabetes, and heart disease later in life. Remarkably, these health effects were also observed in the next generation, suggesting that the famine had caused epigenetic changes in the genes of the affected individuals, which were then inherited by their children.

This phenomenon demonstrates that environmental factors can leave epigenetic marks on our DNA, influencing gene expression in ways that significantly impact our health and development. These marks act as a molecular memory of environmental exposures, subtly altering the way genes are expressed without altering the genetic code itself.

Beyond Genetic Determinism

It becomes increasingly clear that the tapestry of life is far too intricate to be woven by genetic threads alone. This journey through the complex interplay of genetics, environment, and development reveals a biological narrative that transcends the simplistic framework of genetic determinism.

Our examination has underscored the pivotal, yet not exclusive, role of genetics in determining traits and influencing behaviors. While genes undoubtedly set the foundational parameters within which biological processes operate, they are not the sole authors of an organism's destiny. The environment, both external and internal to the organism, plays a crucial and often underappreciated role in shaping how these genetic blueprints are realized.

The profound impact of environmental factors and epigenetic mechanisms challenges the deterministic view of genetics. These factors can modify gene expression, sometimes in long-lasting ways that may even transcend generations, as seen in the study of the Dutch Hunger Winter. Such epigenetic modifications highlight the dynamic nature of gene expression, influenced by life experiences, environmental exposures, and developmental processes.

The complexity of biological systems, where multiple genes interact within intricate networks influenced by a myriad of environmental and developmental factors, defies the reductionist approach of genetic determinism. This complexity suggests a model of biology that is more akin to a web of interactions than a linear cause-and-effect relationship.

The insights gained from challenging genetic determinism have profound implications for understanding human diversity and the process of evolution. They encourage a more holistic view of evolution, one that considers genetic changes in the context of ecological dynamics, developmental processes, and the organism's interaction with its environment. This perspective fosters a deeper appreciation for the diversity of life and the evolutionary pathways that have shaped it.

Moving beyond genetic determinism means embracing a more integrated approach to biology, one that acknowledges the complexities and nuances of genetic expression. It calls for a paradigm that appreciates the interplay between genes and the myriad factors that influence their expression. This approach does not diminish the importance of genetics; rather, it enriches our understanding by situating genetics within a broader biological and environmental context.

Transcending the confines of genetic determinism opens up a richer and more nuanced understanding of life. It invites us to view genetics not as the sole determinant of biological fate but as a part of a dynamic, interconnected system where multiple factors contribute to the tapestry of life. This realization is not just a theoretical exercise; it is a fundamental shift in how we understand ourselves and the natural world, offering a more comprehensive and interconnected view of life's complexity.

Implications for Understanding Evolution

Reflecting on the implications of skeptical viewpoints toward genetic determinism significantly impacts our understanding of evolution. The traditional narrative of evolution, heavily reliant on genetic determinism, posits that genetic mutations and natural selection are the primary drivers of evolutionary changes, including speciation, adaptation, and the emergence of complex traits. However, acknowledging the limitations of this perspective invites a reevaluation of these evolutionary processes, indicating potential problems with current evolutionary theories.

Speciation, the process by which new species arise, has traditionally been viewed through the lens of genetic divergence. Genetic determinism suggests that accumulating genetic mutations over time can lead to reproductive isolation and the formation of new species. However, this viewpoint may oversimplify the process. Environmental factors, ecological niches, and even behavioral changes play significant roles in speciation. These elements can induce or facilitate genetic changes, suggesting that speciation is a more complex, multifaceted process than genetic determinism alone can explain.

Adaptation, the process by which organisms become better suited to their environment, is another cornerstone of evolutionary theory traditionally viewed as a result of genetic mutations that confer survival advantages. However, the role of environmental factors and epigenetic mechanisms suggests that adaptations can also occur in response to environmental cues, without changes to the DNA sequence. This understanding complicates the straightforward narrative of adaptation driven solely by genetic changes, highlighting a problem in the simplistic genetic determinism model.

The development of complex traits, such as the eye or the brain, has often been cited as a triumph of gradual genetic evolution. Yet, the interplay of genetic, environmental, and developmental factors in forming these complex structures poses a challenge to the linear model proposed by genetic determinism. The emergence of complex traits likely involves more than just the accumulation of beneficial mutations; it also includes regulatory networks, gene-environment interactions, and possibly non-genetic heritable factors. This complexity points to a problem in the traditional evolutionary narrative that often underestimates the role of non-genetic factors.

Acknowledging the limits of genetic determinism compels a broader, more nuanced view of evolutionary theory. It suggests that evolution is not a simple, linear process governed solely by genetic changes but a dynamic interplay of various factors. This realization can lead to a more comprehensive understanding of evolution, one that incorporates genetic, environmental, epigenetic, and developmental factors. It challenges researchers to look beyond the gene-centric model and consider the broader ecological and organismal contexts in which evolution occurs.

In conclusion, the implications of skeptical viewpoints on genetic determinism reveal problems with traditional evolutionary theories that have predominantly emphasized a gene-centric approach. Recognizing these limitations is crucial for an holistic understanding of life, an understanding that appreciates the complexity and interconnectedness of biological systems.

UNDERSTANDING MACROEVOLUTION AND MICROEVOLUTION - IMPLICATIONS FOR EVOLUTIONARY THEORIES

In our quest to critically examine the underpinnings of evolutionary theory, a pivotal distinction emerges between macroevolution and microevolution. These concepts, often conflated or oversimplified in public discourse, play crucial roles in our understanding of how life on Earth has changed and diversified over time. This section aims to unravel these two facets of evolution, scrutinizing their definitions, the evidence supporting them, and the implications they carry for the broader narrative of evolutionary theory.

Macroevolution refers to large-scale changes that occur over extended periods, often resulting in the emergence of new species, significant changes in life forms, or even the appearance of entirely new biological structures. It represents the broad strokes of evolutionary change, encompassing events that shape life at the level of species and beyond. Macroevolution is often synonymous with the grand narrative of life's history, encompassing phenomena such as the emergence of mammals, the diversification of flowering plants, or the dramatic shifts marked by mass extinctions.

In contrast, microevolution deals with smaller, more immediate changes within populations or species. It involves alterations in allele frequencies over time, driven by mechanisms such as natural selection, genetic drift, mutation, and gene flow. Microevolution is observable, often within human lifetimes, and encompasses changes like the development of antibiotic resistance in bacteria or the variations in beak size among Galapagos finches documented by Darwin.

As we venture into this exploration, it is crucial to maintain the critical and skeptical lens that has characterized our journey thus far. We will revisit the fossil record, not to rehash its evidentiary role, but to critically examine how it is interpreted in the context of macroevolution. We will explore comparative anatomy and genetic analysis, assessing how these lines of evidence are employed to support or challenge macroevolutionary theories, all while aligning with the themes established in earlier chapters of the book.

Moreover, this section will address the common misconceptions and debates surrounding macroevolution and microevolution. It will explore how these concepts are interconnected yet distinct, and the challenges involved in extrapolating microevolutionary processes to explain macroevolutionary patterns. By dissecting these concepts with a critical perspective, we aim to deepen the reader's understanding

131

of evolution, challenging them to reconsider the complexities and nuances that underlie the evolution of life on Earth.

This section is not just an academic exercise in defining terms; it is an invitation to rethink evolutionary theory's scope and limitations. It underscores the importance of critically evaluating how evolutionary changes, both big and small, are understood , and importantly, how it is interpreted within the scientific community and beyond.

Microevolution: Observable and Immediate Changes

Microevolution encompasses the subtle yet significant changes that occur within species or populations over relatively short timeframes. It is the engine driving the variations we can observe within species, and understanding its mechanisms is pivotal to grasping the broader concepts of evolution.

Mutations, or changes in the DNA sequence, are fundamental to microevolution. They introduce new genetic material into a population, which can lead to changes in traits. While many mutations are neutral or harmful, occasionally, a mutation can confer an advantage that may increase an organism's chances of survival and reproduction. The role of mutations in microevolution is clear, but extrapolating this to account for larger macroevolutionary changes, such as the emergence of new species, is a point of debate. Critics argue that the incremental changes mutations provide are insufficient to explain the complexity observed in macroevolution.

Natural selection acts on the variation within a population, favoring traits that enhance survival and reproductive success.

Over time, this can lead to significant shifts in population characteristics. The peppered moth's color variation during the Industrial Revolution is a classic example, where moths with darker coloring were selected for in polluted environments. However, skeptics of macroevolution point out that while natural selection explains these small-scale adaptations well, it does not necessarily demonstrate how entirely new species or complex organs can evolve.

Genetic drift refers to random fluctuations in the frequency of alleles (variants of a gene) within a population. These random changes can have significant effects, especially in small populations, leading to the loss of genetic diversity. While genetic drift can lead to observable changes within a population, its role in driving larger evolutionary changes remains a subject of debate. Critics suggest that random drift alone cannot account for the structured complexity seen in the biological world.

Gene flow occurs when individuals from different populations interbreed, leading to the exchange of genetic information. This can introduce new alleles into a population, increasing genetic diversity. Gene flow can counteract the effects of genetic drift and natural selection within a population, but its impact on larger evolutionary processes, such as the development of new species, is more complex and less straightforward.

Understanding the mechanisms of microevolution is crucial for comprehending how species adapt and change over time. However, when examining evolutionary theory critically, it becomes evident that these micro-scale processes do not straightforwardly extrapolate to explain macroevolutionary phenomena. This realization calls for a deeper exploration of how large-scale evolutionary changes occur, beyond the scope of microevolutionary mechanisms alone.

Examples and Case Studies: Dog Breeding as a Microevolutionary Process

Dog breeding presents a fascinating and accessible example of microevolution in action. Over generations, selective breeding practices have led to the development of a wide array of dog breeds, each with distinct physical and behavioral traits. This process offers a clear illustration of how human intervention can direct microevolutionary changes within a species.

Selective breeding in dogs involves choosing specific individuals with desired traits to produce offspring. Over time, this selective process has accentuated certain characteristics in different breeds, such as size, coat color, temperament, and even specific physical attributes like the shape of ears or the length of the snout.

This intentional selection mirrors natural selection but in a more accelerated and directed manner. The diversity in dog breeds today, from Chihuahuas to Great Danes, demonstrates how microevolution can lead to significant variation within a species.

Dog breeding has also provided insights into the genetic underpinnings of certain traits. Genetic studies on different dog breeds have helped scientists understand how specific genes or genetic combinations contribute to physical and behavioral traits. This research has broader implications for understanding genetic mechanisms in other species as well.

Limitations and Critiques of Microevolution in Dog Breeding

While dog breeding stands as a prime example of microevolution, it also brings to light significant limitations and critiques, particularly when extrapolating these changes to understand macroevolution.

Bounded Variability within a Species:

The diverse array of dog breeds, despite their striking differences in appearance and behavior, all belong to the same species, Canis familiaris. This fact poses an intriguing question: why, despite centuries of selective breeding and the emergence of extreme phenotypic variations, have these changes not transcended the species boundary? This observation is a key point of critique by those skeptical of macroevolution. It suggests that while selective breeding can lead to substantial variation, it operates within certain biological constraints that prevent the emergence of new species.

The Scale and Nature of Evolutionary Changes:

Critics argue that the changes observed in dog breeds, though visually dramatic, are largely superficial and do not represent the kind of deep, structural changes one might expect in macroevolutionary processes. For example, the differences between a Chihuahua and a German Shepherd are less about fundamental biological divergence and more about variations within a common canine framework. This perspective challenges the idea that the degree and scope of changes observed through microevolution in dog breeding can be directly applied to understanding the processes that drive the evolution of entirely new species or complex biological systems.

Genetic and Developmental Constraints:

Another critique focuses on genetic and developmental constraints. Dogs, regardless of breed, share a common genetic and developmental blueprint that limits how far variation can go. These constraints are thought to be a significant factor in why new species have not emerged from dog breeding. This raises questions about the limits of microevolution and the additional mechanisms or conditions needed for macroevolutionary changes, such as speciation or the development of novel biological structures.

Implications for Evolutionary Theory:

The case study of dog breeding, therefore, presents a more nuanced picture. It is a testament to the power of microevolutionary forces in creating diversity within a species. Yet, it also underscores the challenges in extending these observations to macroevolution. The lack of new species emerging from dog breeds serves as a critique against the assumption that microevolutionary processes alone are sufficient to explain macroevolutionary phenomena. It suggests a more complex interplay of genetic, environmental, and possibly unknown factors in driving the larger scale changes observed in the history of life on Earth.

Macroevolution: Large-Scale Evolutionary Changes

Understanding Macroevolution

Macroevolution stands as a cornerstone concept in evolutionary biology, encompassing the broad-scale changes that occur over extended geological timescales and often result in the emergence of new species, significant alterations in life forms, or even the development of novel biological structures. It is a phenomenon that goes beyond the incremental changes observed in microevolution, delving into the deeper, more transformative processes of life's history.

Macroevolution refers to evolutionary changes that are sufficiently large to result in the differentiation of species, genera, families, or even larger groups. These changes often involve the divergence of life forms from a common ancestor and the emergence of new biological characteristics. Macroevolutionary processes can lead to the development of new species (speciation), the extinction of species, and the evolution of major new features or body plans.

The concept of macroevolution is significant in understanding the overarching patterns and trends in the history of life on Earth. It addresses questions about the origins and fates of major groups of organisms, the dynamics of mass extinctions and subsequent radiations, and the mechanisms that drive the evolution of complex biological systems. Macroevolutionary studies seek to explain not just how organisms adapt and change within specific environments, but how new forms of life emerge and evolve over millions of years.

While microevolution involves changes within species or small groups over relatively short periods, macroevolution encompasses more profound and encompassing changes. It is concerned with the accumulation of microevolutionary changes over vast timescales, leading to the formation of new taxonomic groups. This distinction is crucial in evolutionary biology, as the mechanisms and patterns observed at the macroevolutionary level may differ fundamentally from those at the microevolutionary level. For instance, while natural selection and genetic drift can explain variations within a species, macroevolution involves additional factors such as developmental biology, geographic and ecological constraints, and historical contingencies.

Understanding macroevolution is essential for constructing a comprehensive narrative of life's evolution. It provides a framework for interpreting the fossil record, understanding the relationships between different groups of organisms, and explaining the large-scale patterns and trends observed in the diversity of life. Macroevolutionary research helps to decipher the grand narrative of evolution, piecing together the complex puzzle of how simple ancestral forms evolved into the vast array of life we see today.

Macroevolution is a key concept in evolutionary biology that provides insight into the large-scale changes and processes that shape the diversity of life on Earth. It extends beyond the scope of microevolution, offering a broader perspective on the evolutionary process and addressing some of the most fundamental questions about the origins, diversification, and development of life.

Evidence Supporting Macroevolution

The evidence supporting macroevolution is diverse, encompassing fossil records, comparative anatomy, and genetic analysis. Each of these areas offers insights into the large-scale changes that characterize macroevolution, but they must be critically examined in light of the challenges and limitations discussed in earlier chapters of the book.

Fossil Records: Windows into the Past

The fossil record is often cited as a primary source of evidence for macroevolution. It provides snapshots of past life forms and hints at the evolutionary transitions that have occurred over millions of years. For instance, the progression from early aquatic fish to terrestrial tetrapods is documented with a series of intermediate forms. However, as previously discussed, the fossil record is incomplete and subject to interpretational biases. Gaps in the record, such as the absence of clear transitional forms for some major evolutionary changes, present challenges in using fossils to construct a complete narrative of macroevolutionary processes.

Comparative Anatomy: Tracing Evolutionary Relationships

Comparative anatomy involves examining the structural similarities and differences among living and extinct organisms. Homologous structures, such as the limb bones of humans, whales, and bats, suggest a common ancestral origin despite their different functions. These anatomical similarities provide evidence for macroevolutionary change. However, this approach requires careful interpretation, as convergent evolution can lead to similar features in unrelated groups, complicating the understanding of evolutionary relationships.

Genetic Analysis: Deciphering Evolutionary Lineages

Advances in genetic analysis have significantly enhanced our understanding of macroevolution. By comparing genetic sequences across different species, scientists can infer evolutionary relationships and timescales. Molecular clock techniques, for example, estimate the time of divergence between species based on genetic differences. However, these methods rely on certain assumptions about mutation rates and may not always accurately reflect the complex dynamics of genome evolution. Additionally, the genetic evidence must be reconciled with paleontological data, which can sometimes lead to conflicting interpretations.

Interpreting Evidence with Caution

While fossil records, comparative anatomy, and genetic analysis provide compelling evidence for macroevolution, each source of evidence has its limitations and requires careful, critical interpretation. These forms of evidence, when viewed collectively and scrutinized in light of the challenges discussed in earlier chapters, offer a more nuanced and complex picture of macroevolution. It is through this critical lens that we can gain a deeper understanding of the broad patterns and processes that have shaped the diversity of life on Earth.

Challenges and Debates Surrounding Macroevolution

Macroevolution, the process of significant evolutionary changes that occur at the level of species and above, stands at the forefront of evolutionary biology, yet it is enshrouded in a myriad of challenges and debates. One of the most contentious issues surrounding macroevolution is its observability. Unlike microevolution, which can be observed and quantified within relatively short time frames, macroevolution spans across vast geological periods, often beyond the realm of direct observation. This lack of direct observability necessitates reliance on inference from available evidence, such as fossil records and genetic data. However, this inferential approach to understanding macroevolution has been a point of contention, with skeptics questioning the extent to which such evidence can conclusively demonstrate large-scale evolutionary processes, especially in light of gaps and inconsistencies in the fossil record.

This reliance on inference from available evidence, such as the fossil record and genetic data, has led to debates about the interpretation and conclusiveness of these findings. Skeptics argue that the inferential nature of macroevolutionary evidence can lead to speculative conclusions, particularly in the absence of continuous and comprehensive fossil records.

Another significant debate is whether microevolutionary processes, such as natural selection and genetic drift, are sufficient to account for the large-scale changes observed in macroevolution. While these microevolutionary mechanisms undoubtedly play a role in evolutionary changes, critics question whether they can adequately explain the emergence of entirely new species or complex biological structures. This debate touches on the core of evolutionary theory, challenging whether the accumulation of small-scale changes can lead to the profound transformations implied in macroevolution.

The complexity of biological systems and the emergence of novel traits or species present another challenge to macroevolutionary explanations. How complex structures, such as the eye or the brain, evolved from simpler forms remains a topic of intense study and debate. Critics of macroevolution often point to the concept of "irreducible complexity," arguing that certain biological systems are too complex to have evolved through a series of incremental changes.

Reconciling genetic evidence with paleontological data is another area of contention. While molecular biology has provided new insights into the relationships and divergence times of species, these findings sometimes clash with the evidence from the fossil record. This discrepancy raises questions about the accuracy and interpretation of both genetic and fossil data in reconstructing the history of life on Earth.

In conclusion, the challenges and debates surrounding macroevolution underscore the dynamic nature of evolutionary biology. While macroevolution remains a key concept in understanding the broad patterns of life's history, these debates highlight the need for continuous examination, critical analysis, and integration of multiple lines of evidence. The field of macroevolution is evolving, with new discoveries and methodologies continually shaping our understanding of life's grand narrative.

Implications for Evolutionary Theory

The exploration of macroevolution within the framework of evolutionary theory, while enlightening, also unveils a landscape rife with assumptions, challenges, and interpretative complexities. This perspective not only broadens our understanding of evolution but also casts a critical light on some of the foundational aspects of evolutionary thought.

One of the most significant implications of studying macroevolution is the realization of the substantial inferential leaps often required in evolutionary biology. Given the nature of macroevolutionary changes—occurring over vast spans of time and often without direct observational evidence—much of the theory relies on connecting disparate dots, from fossil fragments to genetic sequences. This reliance on inference leads to a range of interpretations, each with its own set of assumptions and conjectures. Such diversity in interpretation, while a testament to the dynamic nature of scientific inquiry, also highlights the inherent uncertainties and speculative elements within macroevolutionary theories.

Critically, the extrapolation of microevolutionary processes to explain macroevolutionary phenomena is a contentious issue. While microevolutionary changes are observable and well-documented, their role in driving the emergence of new species, novel biological structures, or significant evolutionary shifts remains a subject of intense debate. Skeptics argue that the gradual and incremental changes observed in microevolution may not necessarily account for the more profound transformations implied in macroevolution. This gap in explanation raises questions about the sufficiency of current evolutionary models and suggests the potential need for additional mechanisms or processes to fully comprehend macroevolutionary changes.

Furthermore, the interpretation of the fossil record—a key pillar in studying macroevolution—comes with its challenges. The incompleteness of the fossil record, the potential for misinterpretation of fossilized remains, and the difficulties in accurately dating these finds all contribute to a narrative of macroevolution that is as much about educated guesswork as it is about empirical evidence. The critical examination of these fossils, their contextualization, and the narrative constructed from them can vary significantly, leading to diverse and sometimes conflicting views on the history and process of evolution.

In essence, a critical examination of macroevolution in the context of evolutionary theory reveals a complex and often ambiguous picture. It challenges us to consider the limitations of current scientific methodologies and the assumptions underpinning evolutionary narratives. While macroevolution remains a key concept in understanding the history of life, this critical perspective invites a more cautious and questioning approach to evolutionary explanations, recognizing the complexities, gaps, and ongoing debates that characterize this field of study.

As we conclude our exploration of macroevolution and microevolution in the context of evolutionary theory, it becomes evident that these concepts, while distinct, are intricately linked within the grand tapestry of life's evolutionary narrative. This chapter's critical and skeptical examination of both microevolutionary processes and macroevolutionary changes challenges us to reconsider our understanding of evolution, offering new insights while highlighting significant gaps and questions that remain.

The integration of insights from both microevolution and macroevolution underscores the complexity of evolutionary processes. It reveals that evolution is not a singular, linear narrative but a multifaceted story woven from a diverse array of biological, ecological, and temporal threads. Recognizing the limitations and critiques of both microevolutionary mechanisms and macroevolutionary patterns suggests that our current models of evolution, while robust in many respects, may not fully capture the nuances and complexities of life's evolutionary journey. This realization points to the need for more comprehensive and integrative theories that can accommodate the broad spectrum of evolutionary phenomena.

The role of skepticism and critical analysis in this exploration cannot be overstated. It is through questioning established narratives and rigorously examining evidence that science advances. In evolutionary biology, this means continually testing hypotheses, refining theories, and remaining open to revising our understanding in light of new evidence and perspectives. Such an approach not only strengthens the scientific endeavor but also enriches our appreciation of the natural world and its history.

Looking ahead, the future of evolutionary research appears poised for exciting developments. The questions and challenges raised in this chapter highlight areas ripe for further investigation. Advances in genetic analysis, paleontological discoveries, and a deeper understanding of ecological and environmental influences all hold the potential to shed new light on the mechanisms of evolution. Moreover, interdisciplinary research that bridges genetics, paleontology, ecology, and other fields will be crucial in developing a more holistic understanding of how life evolves.

In sum, this book's examination of macroevolution and microevolution, with its emphasis on critical analysis and skepticism, invites us to view evolution not as a solved puzzle but as an ongoing inquiry. It encourages us to embrace the complexities, engage with the debates, and contribute to the evolving narrative of life's extraordinary evolutionary saga.

Genetic Complexities in Macroevolution: A Critical Perspective

145

We have encountered a myriad of concepts that challenge our understanding of life's history on Earth. Now, we delve into a domain rife with both wonder and controversy — the genetic complexities inherent in the process of macroevolution. This segment of our exploration is dedicated to unraveling the enigmatic tapestry of how significant genetic changes contribute to the grand narrative of life's evolution, particularly at scales transcending individual species.

Macroevolution, a term that evokes images of life's grand transitions — from the emergence of complex multicellular organisms to the rise of major taxonomic groups — is a process marked not only by its outcome but also by its intricate genetic mechanisms. Here, we confront a pivotal question: How do significant additions and alterations in genetic material pave the path for the profound diversity we observe in the natural world? This question is not merely of academic interest; it touches the very core of our understanding of evolutionary processes.

As we embark on this inquiry, it's crucial to approach the subject with a balanced perspective, one that is open to both established scientific understanding and critical skepticism. The field of genetics, with its ever-expanding scope, has unveiled mechanisms such as gene duplication, horizontal gene transfer, and chromosomal rearrangements, each playing a role in shaping the course of evolutionary history. However, these revelations are not without their complexities and conundrums.

In the spirit of scientific rigor and curiosity, we will critically examine these genetic phenomena, probing into how they align with, or challenge, our current evolutionary paradigms. From scrutinizing the role of polyploidy in plant speciation to

146

questioning the incremental nature of complex trait development, we aim to shed light on the more contentious and less understood aspects of macroevolutionary genetics.

This exploration is not intended to debunk or diminish the significance of evolutionary theory; rather, it seeks to enrich our understanding by confronting the challenges and unresolved questions that continue to animate scientists and skeptics alike. Through this critical lens, we aspire to gain a deeper, more nuanced appreciation of the genetic forces that have sculpted the vast and varied tapestry of life on our planet.

Mechanisms and Challenges of Additions of Genetic Information

Gene duplication, a pivotal mechanism in the evolutionary process, has long intrigued biologists and skeptics alike. It serves as a prime example of how new genetic material can arise and significantly impact an organism's evolutionary trajectory. However, while gene duplication's role in adding genetic information is well-documented, it also presents a series of challenges and questions that warrant a more critical examination.

At its core, gene duplication is a process where an organism's genome acquires an additional copy of a gene. This event can happen during DNA replication or as a result of chromosomal rearrangements. The significance of gene duplication lies in its potential to contribute to evolutionary novelty. Once a gene is duplicated, the redundancy it creates allows one copy to maintain its original function, while the other is free to mutate and possibly acquire new functions, a process known as neofunctionalization.

The evolutionary implications of this process are profound. Gene duplication has been linked to the development of new metabolic pathways, adaptations to environmental changes, and even the emergence of complex physiological structures. For instance, in the human genome, families of genes that have expanded through duplication play essential roles in diverse functions like immune response and sensory perception.

However, the evolutionary path from gene duplication to the emergence of new, beneficial traits is not straightforward and is subject to critical scrutiny. Skeptics of macroevolution often point out that the likelihood of a duplicated gene developing a new, advantageous function is overshadowed by the possibility of non-functional or deleterious mutations. This skepticism stems from an understanding of the random nature of mutations and the intricate balance required for new gene functions to integrate seamlessly into an organism's existing biological systems.

Moreover, the initial redundancy created by gene duplication can lead to genetic instability. This redundancy, while providing a buffer against harmful mutations, can also result in maladaptive traits if the duplicated genes are not properly regulated. The integration of new gene functions into complex biological systems requires intricate regulatory adjustments, a process that skeptics argue is not easily achieved through random mutations and natural selection alone.

Critics also question the timescales required for the beneficial integration of duplicated genes. The process of a gene duplication leading to a new functional gene through random mutations might require an extensive period, during which the gene might become non-functional or be lost. This aspect raises questions about the frequency and efficiency of neofunctionalization as a driver of evolutionary change.

In light of these considerations, gene duplication, while undeniably a key player in the story of evolution, is also a topic of ongoing debate and skepticism. Its role in adding genetic information and contributing to evolutionary complexity is a testament to the dynamic nature of genomes. Yet, it simultaneously highlights the intricacies and uncertainties inherent in the evolutionary process, reminding us that the journey from genetic change to evolutionary innovation is often complex and not entirely predictable.

Integration of New Genetic Information

Functional Incorporation: Once a gene is duplicated, the subsequent integration of the new genetic information into the organism's functional genome is not a straightforward process. The new gene must not only acquire a beneficial function but also fit into the existing regulatory networks. This integration requires a series of precise mutations, each contributing to the gene's functional evolution and its regulatory compatibility with the organism's existing genetic framework.

Regulatory Challenges: A key challenge lies in the regulation of the new gene. Genes are not independent entities; they function as part of a complex network, regulated by a multitude of factors including promoters, enhancers, and inhibitors. The duplicated gene must find its place within this network, a process that can be fraught with complications. Improperly regulated genes can lead to detrimental effects, such as uncontrolled cell growth or metabolic imbalances.

Potential Issues and Skeptical Viewpoints

Random Mutations and Functional Integration: Critics often point to the randomness of mutations and question the likelihood of these random events leading to a beneficial and well-integrated new gene. The process requires not just one, but a series of advantageous mutations, each occurring at the right time and in the right genetic context.

Genetic Redundancy and Drift: Immediately following duplication, the new gene often exists as a redundant copy, making it susceptible to genetic drift — random fluctuations in gene frequencies. This redundancy can lead to the new gene's degradation or loss if it does not quickly acquire a beneficial function.

Pace of Evolutionary Change: The integration of new genetic information is a process that unfolds over many generations. Skeptics of rapid evolutionary change argue that the pace at which new genes acquire beneficial functions and integrate successfully into an organism's genome can be much slower than what some evolutionary models suggest.

Evidence from Comparative Genomics: While gene duplication and integration are well-supported by genomic evidence, translating this genetic data into a coherent narrative of functional evolution remains challenging. Comparative genomics provides insights but also reveals the complex and often non-linear paths that genes take in the course of evolution.

In summary, the integration of new genetic information through gene duplication into an organism's existing genome is a process filled with evolutionary potential, yet fraught with challenges. It requires a fine balance of random mutations, regulatory adjustments, and natural selection. This complexity underlines the skepticism surrounding the efficiency and likelihood of new genes contributing to significant evolutionary changes. As such, the integration of new genetic information remains a fertile ground for ongoing research and debate in evolutionary biology.

Chromosomal Changes and Speciation Dynamics

Polyploidy in Speciation

Polyploidy, the condition of having more than two complete sets of chromosomes, is a significant driver of speciation, particularly in the plant kingdom. Its role in macroevolution has been a subject of extensive study and debate, offering insights into how large-scale genetic changes can lead to the emergence of new species.

Mechanism and Occurrence: Polyploidy occurs through events like whole genome duplication or the fusion of gametes with an abnormal number of chromosomes. It is more prevalent in plants due to their often more flexible reproductive strategies. In some cases, polyploidy results from hybridization between different species, leading to allopolyploid organisms with multiple sets of chromosomes from distinct species.

Implications for Speciation: The instant genetic isolation created by polyploidy can lead to rapid speciation. Polyploid organisms, due to their distinct chromosomal makeup, may become reproductively isolated from their diploid ancestors. This isolation is not only a physical barrier but also a genetic one, as hybrids between polyploid and diploid organisms are often sterile or less viable.

Adaptive Advantages: Polyploidy can offer immediate adaptive benefits. Increased genetic material can lead to greater genetic variability and robustness, allowing polyploid organisms to adapt to a wider range of environmental conditions. In some cases, polyploidy is associated with increased size, faster growth rates, or greater resistance to environmental stressors.

Skeptical Viewpoints: Critics often point to the complexity of successful polyploidization. The process involves synchronizing a doubled genome, which can initially lead to developmental instabilities and reduced fertility. The likelihood of successful polyploidy and its contribution to biodiversity is a subject of ongoing debate.

Chromosomal Rearrangements

Chromosomal rearrangements like inversions, translocations, and fusions are other mechanisms contributing to macroevolution and speciation. These rearrangements can have profound effects on genetic compatibility and the evolutionary trajectory of populations.

Effects on Genetic Compatibility: Chromosomal rearrangements can lead to reproductive barriers. For instance, individuals with different chromosomal arrangements may produce offspring with reduced fertility, as seen in cases of chromosomal inversions where gene sequences are flipped within a chromosome.

Role in Speciation: These rearrangements can contribute to the genetic isolation necessary for speciation. Over time, populations with different chromosomal arrangements may diverge significantly, leading to the emergence of new species. This is particularly evident in cases where chromosomal changes are associated with adaptations to specific ecological niches.

Challenges and Controversies: Despite the clear evidence of chromosomal rearrangements in speciation, there are challenges in understanding the full extent of their impact. Critics question how often such rearrangements lead to successful speciation events, given the potential for reduced fertility and viability in offspring. Additionally, the role of chromosomal rearrangements in speciation compared to other genetic changes is a matter of ongoing research and discussion.

In summary, both polyploidy and chromosomal rearrangements play significant roles in the dynamics of speciation and macroevolution. Their impacts on genetic compatibility, reproductive isolation, and adaptive potential are well-documented, yet they also present a complex array of challenges and questions, particularly regarding their frequency, success rate, and overall contribution to the biodiversity we observe in nature.

Skeptical Examination of Genetic Complexity

Incremental Complexity Under Scrutiny

The concept of incremental complexity in evolution, which posits that complex structures and systems evolve through the gradual accumulation of small genetic changes, is a cornerstone of evolutionary theory. However, this concept faces scrutiny and skepticism, particularly when considering the intricate complexity of many biological structures.

» Challenges to Gradualism: Critics of the gradual accumulation model argue that the development of highly complex structures, such as the eye or the brain, through a series of small, incremental steps seems statistically improbable and biologically implausible. They question how random mutations and natural selection could lead to the precise coordination and integration necessary for such complex systems to function.

» The Issue of Intermediate Forms: One of the main points of contention is the existence and viability of intermediate forms. Skeptics ask what advantage would be conferred by the incomplete or partially formed stages of a complex structure. For instance, what would be the function and advantage of a partially formed wing or a partially developed eye?

» Irreducible Complexity: This concept, often cited by proponents of intelligent design, argues that certain biological systems are too complex to have evolved from simpler predecessors. They claim that these systems consist of multiple interdependent parts, all of which must be present simultaneously for the system to function, thus rendering the step-by-step evolutionary process implausible.

» Fossil Record Gaps: Skeptics also point to the gaps in the fossil record, arguing that the lack of transitional forms challenges the idea of gradual complexity. They assert that if incremental changes were the norm, the fossil record should be replete with examples of these intermediate stages.

Questioning Genetic Information Increase

The increase in genetic complexity over time is another aspect of macroevolution that invites skepticism. Critics question the mechanisms by which genetic information increases in complexity and the likelihood of such increases leading to advanced biological functions.

» Random Mutations and Complexity: The role of random mutations in creating new genetic information is a major point of skepticism. Critics argue that the probability of random mutations leading to beneficial, complex genetic sequences is exceedingly low. They often assert that mutations are more likely to be neutral or deleterious rather than beneficial.

» Genetic Entropy: Some skeptics invoke the concept of genetic entropy, suggesting that over time, genetic systems are more likely to degrade than improve. They propose that the accumulation of small, harmful mutations will, in the long run, lead to a decrease in overall genetic health and complexity.

» Computational Models of Evolution: Computational simulations of evolution are sometimes cited to challenge the increase in complexity. Critics argue that these models fail to realistically replicate the complexity of natural evolutionary processes and often require fine-tuning or guidance to achieve desired outcomes, thereby questioning the unguided nature of evolutionary complexity.

In summary, the skeptical examination of genetic complexity in evolution brings to light significant questions about the mechanisms, feasibility, and evidence for the gradual development of complex structures and the increase in genetic information. These critiques emphasize the need for a deeper, more nuanced understanding of evolutionary processes and highlight the areas where further research and exploration are essential.

Genetic Mutations

Genetic mutations, inherently woven into the fabric of life, are alterations in the genetic material of an organism. They are pivotal in the study of evolutionary biology, serving as a primary source of genetic variation, which is a cornerstone of evolutionary change. This introduction aims to unfold the intricate nature of genetic mutations, their types, and their fundamental role in the biological world.

At its core, a genetic mutation is a change in the DNA sequence within a gene or chromosome of an organism. These changes can range from a single nucleotide variation, known as a point mutation, to larger scale alterations such as insertions, deletions, and chromosomal rearrangements. Each type of mutation carries its own potential impact on the organism's genetic makeup and, consequently, its phenotype.

Point mutations, the simplest form, involve a change in a single nucleotide base in the DNA sequence. These can be further categorized into silent mutations, which do not alter the protein produced by a gene, missense mutations, which change a single amino acid in a protein, and nonsense mutations, which create a premature stop codon, potentially leading to a nonfunctional protein.

Beyond point mutations, frameshift mutations, caused by insertions or deletions of nucleotides, can have a more dramatic effect. They alter the reading frame of the genetic code, often resulting in entirely different and nonfunctional proteins. These can significantly impact an organism's development and function.

Chromosomal alterations involve changes in the structure or number of entire chromosomes. These can occur through processes such as inversion, translocation, duplication, or nondisjunction events during cell division. Such mutations can lead to significant genetic disorders and are a key area of study in genetics and evolutionary biology.

Mutations occur through various mechanisms. They can be spontaneous, resulting from errors in DNA replication, or induced by external factors like environmental chemicals, radiation, or even certain biological agents like viruses. While cells have intricate repair mechanisms to correct DNA errors, these systems are not foolproof, and uncorrected errors can lead to permanent mutations.

In evolutionary terms, mutations are the raw material for natural selection. They introduce genetic diversity into populations, upon which selective forces can act, driving the evolution of species. However, the role of mutations in evolutionary change is complex and multifaceted. It encompasses not only the creation of genetic variation but also the intricate interplay of these variations within the ecological and developmental context of an organism.

Thus, genetic mutations, in their myriad forms and impacts, represent a fundamental aspect of life's evolutionary narrative. Their study not only enriches our understanding of biology but also raises critical questions about the mechanisms and dynamics of evolutionary change.

Mechanisms of Mutation

Mutations, the changes in the DNA sequence of an organism, are fundamental to the process of evolution. They occur through various mechanisms, each contributing to the genetic diversity that is crucial for the adaptation and evolution of species. Understanding how mutations occur is essential to comprehend their role in evolutionary biology.

Spontaneous Mutations: Errors in DNA Replication

Spontaneous mutations are a natural part of the cell cycle, occurring randomly during the process of DNA replication. DNA replication is a highly accurate process, but it is not infallible. Errors can happen when DNA polymerase, the enzyme responsible for copying the DNA, incorporates the wrong nucleotide or skips a nucleotide altogether. While the cell has sophisticated repair mechanisms to correct these mistakes, some errors still evade repair and become permanent changes in the DNA sequence.

These spontaneous mutations are random and can occur in any part of the genome. They may lead to a variety of outcomes, from having no discernible effect to causing significant changes in the organism's physiology or behavior. The randomness of spontaneous mutations is one of the critical factors that drive genetic variation within populations.

Induced Mutations: Environmental Influences

In addition to spontaneous mutations, the environment plays a significant role in inducing mutations. Induced mutations are those caused by external factors, including both natural and human-made elements. These factors, known as mutagens, include a wide range of physical, chemical, and biological agents.

Physical mutagens, such as ultraviolet (UV) light and ionizing radiation (e.g., X-rays, gamma rays), cause mutations by damaging the DNA structure. UV light, for instance, can lead to the formation of thymine dimers, where adjacent thymine bases bond together, disrupting the normal base-pairing during DNA replication. Ionizing radiation can break the DNA strands, leading to deletions, insertions, or rearrangements of the genetic material.

Chemical mutagens include a variety of substances, from natural compounds to synthetic chemicals. These can interact with DNA in multiple ways, such as causing base substitutions or intercalating between DNA bases, which can disrupt the replication process. Examples include certain types of industrial chemicals, tobacco smoke, and even some natural plant compounds.

Biological agents such as viruses can also induce mutations. Some viruses integrate their genetic material into the host's genome, which can disrupt normal gene function or regulation. This viral integration can lead to mutations either directly, through the insertion of viral DNA, or indirectly, by causing genomic instability.

The mechanisms of mutation—both spontaneous and induced—highlight the dynamic nature of the genome. They play a pivotal role in generating the genetic diversity upon which evolutionary forces act. However, the occurrence and effects of these mutations are subject to various factors, including the organism's ability to repair DNA damage, the type of mutagen involved, and the specific site of mutation within the genome. Understanding these mechanisms provides crucial insights into the processes that drive evolutionary change, albeit with a complexity and unpredictability that challenge simplistic interpretations of evolutionary dynamics.

Mutations and Evolutionary Change

In the panorama of evolutionary biology, mutations play a critical role as the raw material for evolutionary change. They introduce genetic variations that are the keystones of evolutionary processes. The interplay of these mutations within populations lays the groundwork for the diversity of life forms observed over time.

Mutations manifest in various forms, ranging from neutral to deleterious and occasionally beneficial. Neutral mutations, which neither benefit nor harm an organism, might accumulate over time, subtly shifting a species' genetic landscape. Deleterious mutations, on the other hand, can be detrimental to an organism's fitness, often being weeded out by natural selection. However, it is the beneficial mutations, albeit rare, that are of prime interest in the study of evolution. These mutations confer an advantage to the organism, potentially improving its survival and reproductive success.

The impact of mutations on an organism's fitness is a complex interplay of genetic and environmental factors. A mutation that is advantageous in one environmental context may be neutral or even harmful in another. This dynamic relationship underscores the multifaceted nature of evolutionary change driven by mutations.

Despite their critical role, the assumption that mutations alone can drive significant evolutionary changes is a topic of skepticism. The rarity of beneficial mutations raises questions about their efficacy in shaping complex evolutionary processes. Critics argue that beneficial mutations are too infrequent to account for the vast diversity and complexity observed in the biological world. This perspective challenges the conventional view that incremental genetic changes are sufficient to drive the large-scale transformations seen in macroevolution.

The impact of mutations on evolutionary change is further complicated by the intricate relationship between genotype and phenotype. Linking specific mutations to significant evolutionary developments involves a myriad of factors beyond mere genetic changes. Environmental influences, epigenetic factors, and the organism's developmental context all play roles in how genetic variations manifest as observable traits.

Moreover, the interpretation of mutational impact is fraught with challenges. Distinguishing between causative and correlative relationships in mutation-driven evolution is not straightforward. A mutation associated with a particular trait does not necessarily mean it is the cause of that trait. This complexity adds another layer of difficulty in understanding the role of mutations in evolution.

In light of these challenges, alternative perspectives on the role of mutations in evolution have emerged. Some theories emphasize the importance of environmental and epigenetic factors, suggesting a more holistic approach to understanding evolutionary change. Debates continue in the scientific community about the extent to which mutations contribute to macroevolutionary changes, reflecting the ongoing dynamic nature of evolutionary research.

While mutations are undeniably crucial in evolutionary biology, their role in driving evolutionary change is more nuanced than often portrayed. A comprehensive understanding of evolutionary processes necessitates integrating genetic changes with environmental, developmental, and epigenetic factors. This integrative approach offers a more complete picture of how mutations contribute to the tapestry of life's evolution.

Skeptical Analysis of Mutation-Driven Evolution

In the realm of evolutionary biology, the role of mutations as drivers of significant evolutionary changes has been a subject of extensive study and debate. While mutations are undoubtedly a source of genetic variation, the assumption that they alone are sufficient to drive major evolutionary changes warrants a critical examination.

The core of this skepticism lies in the nature and frequency of mutations that are beneficial. Beneficial mutations, those that confer an advantage to the organism in its environment, are a rarity in the vast landscape of genetic changes. Most mutations are either neutral, having no significant effect on the organism's fitness, or deleterious, potentially harming the organism's chances of survival and reproduction. The infrequency of beneficial mutations poses a significant challenge to the theory that mutations are the primary drivers of major evolutionary advances.

This rarity brings into question the efficacy of natural selection in shaping complex biological traits and systems. Natural selection, often described as the guiding force of evolution, acts on the variations provided by mutations. However, if beneficial mutations are scarce, the scope of natural selection in sculpting intricate biological features and driving the emergence of new species becomes a topic of contention. Critics argue that the incremental and random nature of beneficial mutations seems inadequate to account for the complexity and diversity of life forms that have evolved over millions of years.

Furthermore, the implications of this rarity for the theory of adaptation are profound. Adaptations, traits that increase an organism's ability to survive and reproduce, are traditionally viewed as products of natural selection acting on beneficial mutations. However, the sporadic occurrence of such mutations suggests a more complex picture of how adaptations arise and are maintained within populations.

This critical perspective calls for a broader understanding of evolutionary mechanisms. It suggests that factors beyond genetic mutations – such as genetic recombination, gene flow, epigenetic changes, and environmental influences – might play more significant roles in evolutionary processes than previously acknowledged. This view encourages a reevaluation of the traditional mutation-driven narrative of evolution, proposing a more multifaceted and dynamic approach to understanding how life evolves.

In sum, a skeptical analysis of mutation-driven evolution reveals a complex interplay of genetic and non-genetic factors in the evolutionary process. It highlights the need for continued research and open-mindedness in exploring the mechanisms underlying the evolution of life, acknowledging the limitations of current theories and the ever-evolving nature of scientific understanding.

Challenges in Interpreting Mutational Impact

In examining the role of genetic mutations within the framework of evolutionary theory, we encounter significant challenges in interpreting their impact. The task of linking specific mutations to significant evolutionary changes is fraught with complexities, inviting a more critical and skeptical approach towards traditional evolutionary narratives.

One of the primary difficulties lies in establishing a clear cause-and-effect relationship between particular mutations and major evolutionary developments. While genetic mutations are the source of variation upon which natural selection acts, pinpointing a direct causal link between a specific mutation and a significant evolutionary change is often elusive. This challenge is compounded by the fact that most mutations have subtle effects, and their influence on an organism's phenotype can be heavily moderated by a host of other genetic and environmental factors.

Furthermore, the process of evolution is characterized by a complex interplay of numerous mutations over long periods. This makes it challenging to single out individual mutations as the definitive drivers of major evolutionary shifts. The emergence of new traits or species is typically the result of a cumulative process involving multiple genetic changes, each contributing incrementally to the overall evolutionary trajectory. Disentangling this intricate web to identify the specific role of individual mutations is a task that often goes beyond the capabilities of current scientific methodologies.

Another critical aspect is the distinction between causative (acting as a cause) and correlative relationships (having a mutual relationship) in mutation-driven evolution. Just because a mutation is associated with a particular evolutionary change does not necessarily mean it caused that change. Correlation does not imply causation, and in the realm of genetic mutations, this distinction is particularly important. Many mutations may be linked to certain traits or evolutionary developments, but proving that they are the direct cause of those developments requires rigorous scientific evidence and often remains a matter of debate.

The challenges in interpreting the impact of mutations extend to the realm of macroevolution as well. While mutations are acknowledged as drivers of microevolutionary changes within species, their role in macroevolutionary events such as the emergence of new species or the development of complex biological systems is less clear. The extrapolation of mutation-driven microevolutionary processes to explain macroevolutionary phenomena is an area of ongoing scientific debate and scrutiny.

In conclusion, the challenges in interpreting the impact of genetic mutations on evolutionary changes call for a cautious and critical approach. They highlight the need for a more nuanced understanding of evolution, one that acknowledges the limitations of current theories and remains open to alternative explanations and mechanisms. This skeptical perspective is vital for advancing our understanding of the evolutionary process, encouraging continuous inquiry and exploration in the dynamic field of evolutionary biology.

Evolution in Action - Natural Selection at the Molecular Level

In the dynamic world of molecular biology, the idea of natural selection plays a pivotal role as a driving force of evolutionary change. This foundational concept, central to our understanding of evolution theory, operates at the molecular level, shaping the genetic makeup of populations over time. To grasp the essence of natural selection in this context, it is essential to delve into its mechanisms and implications at the genetic and molecular scales.

At its core, natural selection is a process by which genetic variations that confer an advantage in terms of survival and reproduction tend to increase in frequency within a population, while those that are disadvantageous tend to decrease. These genetic variations arise primarily through mutations – random changes in the DNA sequence of organisms. While many mutations are neutral or even harmful, some can lead to beneficial traits, providing the organism with an edge in its specific environmental context.

In molecular biology, natural selection acts directly on genes and molecules. Every organism carries a vast array of genes, and each gene can exist in different forms, known as alleles. The frequency of these alleles within a population can shift over time under the influence of natural selection. For instance, if a particular allele leads to the production of a protein that enhances an organism's ability to utilize a nutrient, organisms with that allele may have a higher survival rate and reproduce more successfully. Over generations, this allele will become more prevalent in the population.

One of the quintessential examples of natural selection at the molecular level is the evolution of drug resistance in bacteria. When exposed to antibiotics, populations of bacteria may initially be largely susceptible to the drug. However, if a mutation arises that confers resistance to the antibiotic – perhaps by altering a protein target of the drug or by enabling the bacterium to pump the drug out of its cell – bacteria carrying this mutation will survive and reproduce in the presence of the antibiotic. As a result, the resistant allele increases in frequency, and the population evolves to become predominantly drug-resistant.

This process of natural selection at the molecular level is not just a phenomenon observed in microorganisms. It occurs in all living organisms, driving the evolution of diverse traits from metabolic enzymes to immune system components. The molecular changes brought about by natural selection can have far-reaching implications, influencing an organism's physiology, behavior, and ultimately its fitness in the ever-changing tapestry of its environment.

The idea behind natural selection at the molecular level is a fundamental mechanism by which evolutionary change occurs. It acts on the genetic variations within populations, favoring those changes that enhance survival and reproductive success. Understanding this process is key to unraveling the complexities of evolution and the myriad ways in which life adapts and diversifies in response to the challenges and opportunities presented by the environment.

Molecular Evolution and Genetic Variability

Mutations in DNA play a crucial role by contributing to genetic variability, which serves as the raw material upon which natural selection acts. This genetic variability, stemming from mutations, is fundamental to the process of evolution, as it provides the diversity upon which selective forces can operate.

Mutations, as random alterations in the genetic material of an organism, can lead to new genetic variants or alleles. These variations in the genetic code can manifest in various ways, from subtle changes in protein function to more noticeable alterations in physical traits or behaviors. The diversity created by these mutations is critical for a population's ability to adapt and evolve in response to environmental challenges.

However, the impact of these mutations is not uniformly beneficial. While some mutations can confer advantages to the organism, enhancing its survival or reproductive success, others can be neutral or even detrimental. A beneficial mutation might increase an organism's fitness by improving its ability to procure resources, evade predators, or withstand environmental stresses. For example, a mutation that results in a more efficient enzyme could give an organism an edge in metabolizing a nutrient, thereby enhancing its survival prospects.

Conversely, detrimental mutations can decrease an organism's fitness by impairing vital functions or reducing its ability to cope with environmental challenges. Such mutations may lead to malformations, reduced functionality, or increased vulnerability to diseases. These disadvantageous mutations are often weeded out by natural selection, as the organisms carrying them are less likely to survive and reproduce.

The role of neutral mutations in evolution is also significant. These mutations do not have an immediate impact on an organism's fitness but contribute to genetic diversity within a population. Over time, some of these neutral mutations may become advantageous or disadvantageous due to changes in the environment or other evolutionary pressures.

The interplay of these various types of mutations creates a dynamic genetic landscape within populations. Natural selection acts on this landscape, favoring alleles that enhance fitness and disfavoring those that impede it. This ongoing process leads to changes in allele frequencies over generations, driving the evolutionary adaptations observed in populations.

In essence, the concept of molecular evolution and genetic variability underscores the dynamic nature of evolution at the genetic level. It highlights the importance of mutations in providing the genetic diversity necessary for natural selection to operate. However, the randomness of mutations and their varying impacts on fitness also introduce elements of unpredictability and complexity into the evolutionary process. This complexity challenges simplistic views of evolution and necessitates a nuanced understanding of how genetic changes translate into evolutionary adaptations.

Examples of Molecular Evolution

Molecular evolution, a testament to the power of genetic variability and natural selection, is exemplified in numerous cases across the biological world. Two notable examples are the development of antibiotic resistance in bacteria and the changes in the hemoglobin molecule in human populations living at high altitudes. These instances not only illustrate the process of molecular evolution but also highlight its profound implications.

Antibiotic Resistance in Bacteria:

The emergence of antibiotic resistance in bacteria is one of the most striking examples of molecular evolution. This process involves the genetic adaptation of bacterial populations to survive in the presence of antibiotics, drugs that were initially effective in killing or inhibiting their growth. Resistance often arises from mutations in bacterial DNA that confer a survival advantage in an antibiotic-rich environment. These mutations might enable bacteria to degrade the antibiotic, alter the drug's target site, or increase the efflux of the antibiotic out of the bacterial cell.

Over time, as antibiotics are used, the selective pressure they impose on bacterial populations leads to an increase in the frequency of these resistant genes. Bacteria with the advantageous mutations survive and proliferate, while those without them are eliminated. This evolution of antibiotic resistance, a clear example of natural selection at work, has become a significant concern in medicine, as it renders many standard treatments for bacterial infections less effective or even obsolete.

Hemoglobin Adaptations in High Altitude Populations:

Another compelling example of molecular evolution is observed in human populations living at high altitudes, such as in the Tibetan Plateau, the Andes, and the Ethiopian Highlands. In these environments, the lower oxygen levels pose a challenge to survival and normal physiological functioning. Yet, these populations exhibit adaptations that allow them to thrive in such conditions, notably changes in the hemoglobin molecule, which is responsible for transporting oxygen in the blood.

Studies have shown that these populations have evolved genetic variants of hemoglobin that allow for more efficient oxygen uptake and transport under low-oxygen conditions. These adaptations are the result of natural selection acting on genetic mutations that confer a survival advantage in high-altitude environments. This molecular adaptation has enabled these human populations to maintain normal oxygen levels, avoiding complications associated with high-altitude living.

Both these examples, antibiotic resistance in bacteria and hemoglobin adaptations in high-altitude human populations, underscore the profound effects of molecular evolution driven by natural selection. They demonstrate how genetic mutations, when advantageous, can lead to significant changes at the molecular level, enabling organisms to adapt and survive in changing or challenging environments.

Alternative Explanations for Antibiotic Resistance in Bacteria

While the development of antibiotic resistance in bacteria is often cited as a clear case of natural selection acting on random mutations, some alternative theories and perspectives suggest a more complex interplay of factors.

> » **Horizontal Gene Transfer (HGT)**: Beyond mutations, HGT plays a crucial role in spreading antibiotic resistance. Bacteria can acquire resistance genes from other bacteria, not just through vertical

inheritance (from parent to offspring) but also horizontally from unrelated bacteria. This process can rapidly disseminate resistance traits across different bacterial species and populations, suggesting that the spread of antibiotic resistance involves a network of genetic exchanges, not solely mutation-driven evolution.

» **Environmental Stressors and Induced Mutations:** Some researchers propose that environmental stressors, such as the presence of antibiotics, might induce higher mutation rates in bacteria. This adaptive mutation theory suggests that bacteria may have mechanisms to increase genetic variability in response to stress, which could accelerate the development of antibiotic resistance beyond what would be expected from random mutations alone.

Alternative Perspectives on Hemoglobin Adaptation in High Altitude Populations

The adaptation of hemoglobin in high-altitude human populations is a remarkable example of human adaptation. However, alternative explanations provide additional layers of understanding:

» **Epigenetic Adaptation:** Some scientists argue that alongside genetic mutations, epigenetic mechanisms could play a role in high-altitude adaptation. Epigenetic changes, which involve modifications in gene expression without altering the DNA sequence, could rapidly adjust to environmental changes like low oxygen levels, offering a complementary mechanism to genetic adaptation.

» **Pre-existing Genetic Variability:** Another perspective is that high-altitude adaptations may not solely arise from new mutations but could also result from the selection of pre-existing genetic variability within human populations. This variability, present before exposure to high-altitude environments, could provide the raw material for natural selection to act upon, challenging the view that new mutations are the primary drivers of such adaptations.

These alternative theories and explanations add depth to our understanding of molecular evolution. They suggest that evolutionary changes at the molecular level can result from a combination of genetic mutations, gene flow, epigenetic modifications, and environmental influences. This broader view challenges the notion that molecular evolution is driven solely by random mutations and natural selection, proposing a more intricate and interconnected model of evolutionary change.

Natural Selection and Gene Frequencies

Natural selection plays a pivotal role in shaping the genetic composition of populations. It acts as a filter, sifting through the genetic variations produced by mutations, and over time, influences the frequencies of genes within a population. This dynamic process is central to our understanding of how populations evolve and adapt.

173

Natural selection operates on the principle that certain genetic traits confer a survival or reproductive advantage in a specific environment. Genes that lead to advantageous traits tend to increase in frequency within the population because individuals with these traits are more likely to survive and reproduce, passing these genes to their offspring. Conversely, genes that result in deleterious traits decrease in frequency as they negatively impact an organism's fitness, reducing its chances of survival and reproduction.

For instance, a gene variant that enhances an organism's ability to forage for food or evade predators would be considered advantageous. Over generations, natural selection would favor individuals with this variant, leading to an increased frequency of this gene in the population. On the other hand, a gene variant that hinders an organism's ability to perform essential functions would be selected against, leading to a decrease in its frequency.

The Role of Genetic Drift

Genetic drift, a mechanism distinct from natural selection, also plays a role in changing gene frequencies, particularly in small populations. It refers to random fluctuations in the frequency of alleles from one generation to the next. Unlike natural selection, which is a non-random process driven by the environment, genetic drift is stochastic and can lead to the loss or fixation of alleles irrespective of their adaptive value. Genetic drift can significantly impact the genetic makeup of a population, leading to reduced genetic diversity and potentially influencing the population's ability to adapt to changing environments.

Gene Flow and Its Impact

Gene flow, the transfer of alleles from one population to another through migration, is another factor influencing gene frequencies. When individuals from one population interbreed with another, they introduce new genetic variations. This process can counter the effects of genetic drift by introducing new alleles, enhancing genetic diversity, and preventing populations from becoming genetically isolated.

Bottleneck Effects on Genetic Diversity

The bottleneck effect, a specific type of genetic drift, occurs when a population undergoes a significant reduction in size due to environmental events or other pressures. This reduction can cause a drastic change in gene frequencies, often reducing genetic diversity. The surviving population's genetic makeup might not represent the original population's genetic diversity, leading to a loss of alleles and potentially impacting the population's long-term viability.

Examples of Natural Selection at the Molecular Level

Natural selection at the molecular level could be through various examples where environmental pressures might have led to significant molecular changes in organisms. These changes, often reflected in the evolution of specific enzymes, alterations in protein structures, or the development of new metabolic pathways, provide insight into the adaptive nature of molecular evolution.

The Evolution of Lactase Persistence

One notable example is the evolution of lactase persistence in human populations. Lactase is an enzyme necessary for the digestion of lactose, the sugar found in milk. In most mammals, the expression of the lactase gene decreases after weaning, leading to lactose intolerance in adulthood. However, in certain human populations, particularly those with a long history of dairy farming, lactase persistence has evolved, allowing adults to digest lactose.

This adaptation is the result of natural selection acting on genetic mutations in the regulatory region of the lactase gene. These mutations lead to continued lactase production into adulthood. In populations where dairy farming and milk consumption were prevalent, individuals with lactase persistence had a nutritional advantage and were more likely to survive and reproduce, leading to an increase in the frequency of these mutations.

Critique of Lactase Persistence Evolution

While the evolution of lactase persistence is often attributed to natural selection, some critics argue that the story might be more complex. Alternative explanations suggest that cultural practices, such as dairy farming, may have co-evolved with lactase persistence, rather than simply selecting for it. This perspective implies a more intricate interaction between genetic evolution and human cultural practices. Additionally, some researchers question the uniformity of the evolutionary pressures across different populations, suggesting that lactase persistence may have developed under varying conditions and not solely due to dairy consumption.

The Evolution of Antibiotic Resistance in Bacteria

Another classic example is the evolution of antibiotic resistance in bacteria, a direct consequence of natural selection in response to the use of antibiotics. When exposed to antibiotics, bacteria with mutations conferring resistance have a survival advantage. These mutations can involve various molecular changes, such as alterations in the target enzyme that the antibiotic inhibits, changes in cell membrane permeability to prevent antibiotic entry, or the acquisition of genes that degrade or modify the antibiotic.

The rapid rise in antibiotic-resistant bacterial strains exemplifies how natural selection can lead to significant molecular adaptations in a relatively short period. This phenomenon is a critical issue in modern medicine, highlighting the adaptability of bacteria and the evolutionary arms race between microbial evolution and pharmaceutical development.

Alternative Views on Antibiotic Resistance in Bacteria

The evolution of antibiotic resistance is a critical example of natural selection. However, some critics point out that the development of resistance can also be influenced by horizontal gene transfer, which allows for the rapid spread of resistance genes across different bacterial species. This suggests that the process is not just a result of mutation and selection within a single population but also involves gene exchange networks. Furthermore, there are questions about the long-term sustainability of resistance traits when the selective pressure (antibiotic presence) is removed, challenging the idea that these changes are always permanent or beneficial.

Photosynthesis in Plants

A remarkable example of molecular adaptation is the evolution of C4 photosynthesis in certain plant species. C4 photosynthesis is an advanced mechanism that allows plants to efficiently fix carbon dioxide in environments with high temperatures and low carbon dioxide concentrations. This process involves the evolution of specific enzymes and a unique leaf anatomy to concentrate carbon dioxide around the enzyme Rubisco, reducing photorespiration.

The emergence of C4 photosynthesis in some plant lineages, such as certain grasses and members of the Amaranthaceae family, is a result of natural selection favoring plants that could photosynthesize more efficiently under specific environmental conditions. This adaptation represents a complex interplay of genetic, biochemical, and anatomical changes driven by molecular evolution.

Critiques of C4 Photosynthesis Evolution

The evolution of C4 photosynthesis is often cited as a classic case of adaptive evolution. However, some scientists propose that the initial steps in the evolution of C4 photosynthesis might have occurred due to random genetic drift rather than direct selection for increased photosynthetic efficiency. This view suggests that certain pre-adaptive traits might have arisen by chance and only later were co-opted for their current function in C4 photosynthesis. This perspective emphasizes the role of chance and serendipity in evolutionary processes, alongside natural selection.

Skeptical Perspective on Molecular Complexity

The conventional narrative within evolutionary biology posits natural selection as the key driver of life's molecular complexity and diversity. However, a critical analysis of this perspective reveals several limitations and invites the exploration of alternative explanations.

Limitations of Natural Selection in Explaining Molecular Complexity

Natural selection, while undeniably influential, may not fully account for the intricate complexity observed at the molecular level. The process of natural selection operates on existing genetic variations, selecting traits that confer survival advantages. However, the emergence of complex molecular structures and systems, such as the ribosome, the intricate machinery of cell division, or the complex signaling pathways in cells, poses a challenge to the explanatory power of natural selection alone.

One critical limitation is the concept of "irreducible complexity," which argues that certain biological systems are too complex to have evolved through a series of incremental adaptations. These systems, it is argued, would not function if they were less complex, challenging the idea that they could have developed through the gradual accumulation of beneficial mutations.

Furthermore, the randomness of mutations raises questions about their ability to lead to highly ordered and complex molecular structures. Critics argue that random genetic changes, even when filtered through the lens of natural selection, might be insufficient to explain the highly organized and specialized molecular machinery found in living organisms.

Alternative Explanations and Theories

In light of these challenges, several alternative explanations and theories have been proposed to account for molecular complexity:

>   » **Non-Darwinian Evolutionary Mechanisms:** Some scientists advocate for the role of non-Darwinian mechanisms, such as neutral evolution and genetic drift, in shaping molecular complexity. These mechanisms suggest that some aspects of molecular evolution might be driven more by random chance and historical contingencies than by strict adaptive selection.

>   » **Epigenetic Factors:** Epigenetics, the study of heritable changes in gene expression that do not involve changes to the underlying DNA sequence, offers another perspective. Epigenetic mechanisms can influence the development and function of organisms in ways that are not solely determined by their genetic code, suggesting a more complex interplay between genetics and environment.

>   » **Co-option and Exaptation:** Theories of co-option and exaptation propose that some complex molecular structures may have evolved from simpler precursors initially used for different functions. These structures might have been co-opted for new purposes, gradually acquiring additional complexity.

>   » **Systems Biology Approach:** A systems biology approach emphasizes the role of networks and interactions among genes, proteins, and other molecular entities in generating complexity. This perspective views molecular complexity as arising from

the dynamic interplay of numerous components within biological systems, rather than from individual genes or mutations.

The theory of natural selection, faces significant skepticism when it comes to explaining the origin and complexity of intricate molecular machinery within cells, such as ribosomes and DNA replication mechanisms. These complex systems, essential for the survival and functioning of all living organisms, present a substantial challenge to the idea that natural selection alone can account for their existence.

Complexity of Molecular Machinery: Ribosomes and DNA Replication

The ribosome, a complex molecular machine responsible for protein synthesis, is an example of an incredibly intricate structure composed of RNA and proteins. Its evolution poses a conundrum: the ribosome is essential for producing proteins, yet it is itself made up of proteins. This chicken-and-egg problem raises questions about how such a system could have evolved incrementally through natural selection if its basic functionality requires a level of complexity that seems unattainable in simple, incremental steps.

Similarly, the mechanisms of DNA replication involve a suite of enzymes and proteins working in a coordinated and highly precise manner. The replication machinery must not only copy the genetic material accurately but also repair errors, untangle DNA strands, and ensure the correct distribution of DNA to daughter cells. The complexity of this system, particularly the error-checking and repair mechanisms, poses questions about the feasibility of these features evolving through a series of small, advantageous mutations alone.

Alternative Perspectives: Complexity Beyond Natural Selection

Given these complexities, alternative theories and perspectives have emerged to explain the evolution of such intricate molecular systems:

» **Prebiotic Evolution and Self-Organization:** Some theories propose that basic components of molecular machinery, like ribosomal RNA, may have originated in a prebiotic world through processes of chemical evolution and self-organization. This perspective suggests that the foundations of complex molecular systems could have been established before the advent of life as we know it.

» **Horizontal Gene Transfer and Genetic Fusion:** In the case of complex enzymes and proteins, horizontal gene transfer (the movement of genetic material between unrelated organisms) and genetic fusion events (where separate genetic sequences merge to form new genes) are proposed as mechanisms that could rapidly increase complexity, bypassing the slow pace of mutation and selection.

» **Modular Assembly and Exaptation:** Another perspective is that complex molecular machinery evolved from simpler modular components that initially served different functions. These components could have been co-opted and recombined to form more complex systems, a process known as exaptation.

In conclusion, natural selection and its ability to solely account for the development of complex molecular machinery remains a topic of debate and investigation. Understanding the origins of such intricate systems likely requires an integration of various evolutionary processes and mechanisms, extending beyond the traditional scope of natural selection.

Beyond Genetics - The World of Epigenetics

.In the evolving landscape of biological sciences, epigenetics stands at the forefront of a paradigm shift, challenging conventional understandings of evolution. Epigenetics, broadly defined, involves changes in gene expression that do not alter the DNA sequence itself. These changes are influenced by environmental factors and can affect an organism's traits in significant ways. As the study of epigenetics gains momentum, it presents both an opportunity and a challenge to traditional evolutionary theories.

Defining Epigenetics

Epigenetics refers to a set of mechanisms that alter gene activity without changing the genetic code. These mechanisms include DNA methylation, histone modification, and RNA-associated silencing. They act as molecular switches that can turn genes on or off, and in some cases, these changes can be passed down to subsequent generations. This heritability of epigenetic marks, especially in response to environmental stimuli, introduces a new layer of complexity to the understanding of inheritance and evolution.

Epigenetics and Evolution: A Skeptical Perspective

Traditionally, evolutionary theory has been grounded in the concept of genetic mutation as the primary source of variation for natural selection to act upon. However, epigenetics introduces non-genetic factors into this equation. It suggests that organisms can rapidly adapt to environmental changes through mechanisms that are not directly tied to changes in DNA sequences. This rapid adaptability, potentially inheritable across generations, poses a significant challenge to the classical view of gradual evolution driven solely by random genetic mutations.

Moreover, the concept of epigenetic inheritance revives aspects of Lamarckian evolution – the idea that acquired characteristics can be passed on to offspring. While largely dismissed in the wake of Darwinian theory, the possibility of epigenetic inheritance has reignited discussions about the mechanisms of evolutionary change and the inheritance of traits.

Critiques and Challenges

Critics of traditional evolutionary theory point to epigenetics as evidence of overlooked complexities in how organisms evolve. They argue that evolution is not just a story of genetic change but also one of how organisms interact with their environment and how these interactions can have heritable effects.

Skeptics question the extent to which epigenetic changes can be stable and heritable over multiple generations, thereby influencing evolutionary processes. The degree to which epigenetic changes can lead to new species or significantly alter the course of evolution remains a topic of ongoing debate.

The emergence of epigenetics in the discourse on evolution adds a fascinating layer to the narrative of life's development. It challenges us to rethink traditional notions of heredity and evolution, suggesting a more nuanced interplay between genes, environment, and inheritance. While epigenetics opens new avenues for understanding evolutionary processes, it also invites skepticism and critical inquiry, underscoring the complexity of unraveling the mechanisms that drive the diversity of life.

Epigenetics, has indeed become a focal point for alternative explanations and critiques of traditional evolutionary theories. Here are some key points where epigenetics intersects with critiques of evolution:

» **Inheritance of Acquired Characteristics:** One of the most significant ways in which epigenetics challenges traditional evolutionary theory is by rekindling the idea of the inheritance of acquired characteristics. This concept suggests that traits acquired during an organism's lifetime can be passed on to its offspring. Epigenetic mechanisms, such as DNA methylation or histone modification, can induce changes in gene expression in response to environmental factors, and some of these changes can be transmitted to subsequent generations. This challenges the Neo-Darwinian view, which holds that only genetic changes (mutations) can be inherited and contribute to evolutionary processes.

» **Rapid Adaptive Responses:** Epigenetics provides a mechanism for organisms to adapt rapidly to environmental changes within a single generation, without changes in DNA sequence. This ability for quick adaptation through epigenetic modifications can influence fitness and survival, potentially playing a role in evolution. Critics of traditional evolutionary theory argue that this rapid adaptability has been underestimated and should be integrated into our understanding of evolutionary processes.

» **Non-Genetic Sources of Variation:** Epigenetics introduces non-genetic sources of variation into the evolutionary equation. This challenges the traditional focus on genetic mutations as the sole source of variation for natural selection to act upon. Epigenetic modifications can produce phenotypic variations that are subject to natural selection, adding a layer of complexity to the evolutionary process.

» **Transgenerational Epigenetic Inheritance:** Perhaps the most controversial aspect is the concept of transgenerational epigenetic inheritance, where epigenetic marks established in response to environmental factors are passed down across generations. This phenomenon, if substantiated, would suggest an additional layer of heredity beyond DNA, potentially reshaping our understanding of how traits are inherited and how evolution occurs over generations.

» **Critiques of Epigenetic Contributions to Evolution:** Despite these intriguing possibilities, there are critiques of the extent to which epigenetics can influence evolution. Some scientists caution that the stability of epigenetic modifications across generations is not well established and that the evolutionary significance of these changes remains unclear. There is ongoing debate about the durability and impact of epigenetic changes in comparison to genetic mutations in driving long-term evolutionary changes.

In summary, epigenetics has opened up new avenues for understanding evolutionary processes, offering alternative explanations and challenging some of the traditional tenets of evolutionary theory.

CHAPTER 5

THEORIES OF HUMAN
EVOLUTION

Introduction: The Significance of Human Evolution in Evolutionary Theory

Human evolution stands as one of the most captivating and debated topics within the realm of evolutionary biology. It delves into the origins and development of Homo sapiens, offering insights into our biological and cultural heritage. This chapter aims to explore the multifaceted narrative of human evolution, situating it within the broader context of evolutionary theory while maintaining a critical and questioning lens.

The study of human evolution is significant for several reasons. First, it provides a window into the processes that have shaped not just the human species but life itself. The journey of human evolution, from our early hominid ancestors to modern humans, is a testament to the complex interplay of genetics, environment, and chance that characterizes evolutionary change. It showcases the dynamics of adaptation, speciation, and survival in the face of shifting ecological and climatic landscapes.

Second, human evolution challenges and enriches our understanding of evolutionary theory. It prompts questions about the mechanisms of evolution, such as the role of natural selection in shaping human traits and behaviors. It also raises inquiries about the extent to which our evolutionary past influences our present and future – biologically, socially, and culturally.

However, despite its significance, the narrative of human evolution is not without controversies and complexities. Traditional theories of human evolution, while providing valuable frameworks, have been met with skepticism and critical analysis. Conflicting evidence, alternative interpretations, and ongoing scientific debates highlight the evolving nature of our understanding of human origins.

This chapter will traverse through various aspects of human evolutionary study. It will examine the main theories of human evolution, such as the Out-of-Africa model and Multiregionalism, and critically analyze the evidence supporting these theories. We will delve into the world of hominid fossils, discussing their role in shaping our understanding of human evolution and the controversies surrounding their interpretation. Furthermore, the chapter will explore the use of genetic evidence in tracing human ancestry, evaluating the limitations and assumptions behind these methods.

As we peel back the layers of this fascinating subject, we maintain a skeptical perspective, questioning established narratives and remaining open to new discoveries and interpretations.

Human evolution, a field rich with theories and hypotheses, offers various explanations for how modern humans, Homo sapiens, came to be. Two of the most prominent theories are the "Out-of-Africa" model and the "Multiregionalism" theory, each proposing different mechanisms and timelines for human evolution.

Out-of-Africa Theory

The Out-of-Africa theory, also known as the "Recent African Origin" model, posits that modern humans evolved relatively recently in Africa and then migrated outwards to inhabit other continents. This theory is supported by a range of genetic and fossil evidence suggesting that Homo sapiens appeared in Africa around 200,000 to 300,000 years ago. Proponents of this model argue that as humans migrated, they replaced local archaic human populations, such as Neanderthals in Europe and Denisovans in Asia, without significant interbreeding.

The strength of the Out-of-Africa theory lies in its alignment with the patterns observed in human genetic diversity, which show the highest levels in African populations, indicating a longer period of human habitation. Mitochondrial DNA studies, which trace maternal lineage, and Y-chromosome studies, tracing paternal lineage, both support an African origin for modern humans.

Multiregionalism Theory

In contrast, the Multiregionalism theory proposes that modern humans evolved from earlier human species that had already migrated out of Africa. According to this theory, Homo sapiens emerged simultaneously in several parts of the world, evolving from Homo erectus populations that had left Africa over a million years ago. Proponents of Multiregionalism argue that there was continuous gene flow between geographically separate populations, preventing speciation and resulting in the evolution of modern humans in different regions.

One of the key arguments in favor of Multiregionalism is the presence of regional continuity in some anatomical traits seen in ancient human fossils. This theory also accounts for the possibility of interbreeding between modern humans and archaic human species, as evidenced by genetic data showing Neanderthal and Denisovan DNA in modern non-African human populations.

Skeptical Perspective on Human Evolution Theories

While both the Out-of-Africa and Multiregionalism theories have merits, they also face skepticism and challenges. The Out-of-Africa model's emphasis on a complete replacement of archaic humans by modern humans has been questioned in light of genetic evidence of interbreeding. Similarly, the Multiregionalism theory, while explaining the continuity of regional traits, struggles to reconcile with the genetic evidence that points towards a recent African origin.

The debate between these theories reflects the complexity of human evolution and the difficulty in reconstructing our past. It underscores the need for a nuanced understanding that considers the possibility of overlapping and converging evidence. Recent discoveries and advancements in genetic analysis continue to refine and sometimes challenge these theories, suggesting a more intricate story of human evolution than previously thought.

Critical Key figures and milestones in the study of human evolution

The study of human evolution has seen contributions from figures who have challenged conventional perspectives and brought to light critical aspects that question established theories. Here are some notable figures and milestones with a critical angle on human evolution theories:

» Alfred Russel Wallace (1823-1913): Co-discoverer of the theory of natural selection, Wallace later in life expressed doubts about natural selection's ability to account for human intelligence and consciousness.

» Pierre Teilhard de Chardin (1881-1955): A paleontologist and philosopher, Teilhard de Chardin proposed ideas that combined evolutionary theory with spirituality, suggesting that human evolution is moving towards a more complex and conscious state, a view that stirred controversy in scientific circles.

» Stephen Jay Gould (1941-2002): A prominent paleontologist and evolutionary biologist, Gould was known for his theory of punctuated equilibrium, co-developed with Niles Eldredge. This theory proposed that species undergo rapid changes in relatively short periods, challenging the traditional view of gradual evolution.

» Richard Lewontin (1929-2021): A geneticist known for his work on human variation, Lewontin highlighted the genetic similarity among human populations, questioning the biological basis of race and challenging views on human evolutionary differences.

» Rebecca Cann (1951-present): Cann's work on mitochondrial DNA in the 1980s played a significant role in the Out-of-Africa theory but also opened discussions about the interpretation of genetic evidence in tracing human ancestry, highlighting the complexity and uncertainties in understanding human evolutionary pathways.

Critical Milestones in Human Evolution Research

» 1972: Introduction of the theory of punctuated equilibrium by Gould and Eldredge, challenging the gradualism aspect of Darwinian evolution.

» 1980s: Mitochondrial DNA studies revealing the "Mitochondrial Eve," sparking debates on human origins and migrations, and challenging simplistic interpretations of genetic evidence.

» 1990s-2000s: Increasing discoveries of hominid fossils like Ardipithecus ramidus, challenging the linear progression model of human evolution and suggesting a more complex evolutionary tree.

» 2010: Ancient DNA studies revealing interbreeding between modern humans and Neanderthals, questioning the strict replacement model proposed by the Out-of-Africa theory.

» Recent Years: Advancements in genomics and paleogenetics continually reshaping our understanding of human evolution, highlighting complexities in the interplay between genetics, environment, and culture.

These critical figures and milestones represent a more questioning and nuanced view of human evolution, emphasizing the challenges in unraveling the complex history of human development.

Skeptical View on Theories of Human Evolution

In the quest to understand human origins, mainstream theories of human evolution, while providing significant insights, have not been without their controversies and debates. A critical examination of these theories reveals conflicting evidence and highlights the complexity of piecing together our evolutionary past.

Conflicting Evidence in Human Evolution

The journey of human evolution, as depicted by prevailing theories such as the Out-of-Africa model and Multiregionalism, faces challenges from various conflicting pieces of evidence. One area of contention lies in the fossil record. While the fossil record has been instrumental in shaping our understanding, it is also characterized by gaps and inconsistencies. For instance, the discovery of Homo floresiensis (the "Hobbit") in Indonesia and Homo naledi in South Africa challenges the straightforward narrative of human evolution from Homo erectus to modern humans, suggesting a more complex web of human species coexisting and possibly interacting.

Moreover, advancements in genetic research have both clarified and complicated the picture. Genetic studies have shown interbreeding between Homo sapiens and archaic humans like Neanderthals and Denisovans, blurring the lines between distinct human species. This interbreeding challenges models that strictly separate human species and suggests a more intertwined human ancestry.

Debates Among Scientists

The field of human evolutionary studies is marked by vibrant and ongoing debates among scientists. These debates often revolve around the interpretation of new fossil discoveries, the methodologies used in genetic analysis, and the integration of archaeological findings with genetic data. For instance, the extent to which genetic evidence can inform us about migration patterns and population structures of early humans remains a topic of debate, with some researchers cautioning against over-reliance on genetic data to construct evolutionary narratives.

Another area of debate concerns the role of environmental changes in human evolution. While some scientists emphasize the influence of climatic and ecological shifts in shaping human development, others argue for a greater focus on the social and cultural factors that have also played a significant role.

Critique of Mainstream Theories

The mainstream theories of human evolution, while foundational in our understanding of human origins, encounter substantial critiques when examined through a more critical and comprehensive lens. These critiques not only question the specifics of these theories but also challenge the broader narrative and methodology of how human evolution is portrayed and understood.

> » **Linear and Progressive Narratives:** A significant criticism is directed towards the often linear and progressive narratives portrayed by mainstream

theories. These narratives typically depict human evolution as a straightforward, unidirectional progression from primitive to more advanced forms, culminating in modern Homo sapiens. Critics argue that this portrayal oversimplifies the evolutionary process, reducing it to a series of stepwise improvements. Such a perspective fails to account for the complexity and diversity of human forms throughout history. It overlooks the possibility that different human species, with varying traits and capabilities, may have coexisted and interacted, contributing to a more intricate evolutionary landscape than a simple linear progression.

» **Overemphasis on Certain Aspects:** Mainstream theories are also critiqued for overemphasizing certain aspects of the fossil record while neglecting others. For instance, there tends to be a focus on cranial capacity as a primary indicator of evolutionary advancement, potentially overshadowing other significant aspects such as tool use, cultural practices, or environmental adaptations. This selective emphasis can lead to a skewed understanding of human evolution, one that prioritizes certain traits over others without fully appreciating the multifaceted nature of evolutionary change.

» **Methodological Concerns:** The methodologies employed in studying human evolution are subject to critique as well. For example, the reliance on morphological analysis of fossil remains to draw conclusions about behavioral and cognitive capabilities of early humans is seen as speculative by some researchers. Similarly, the use of molecular clock techniques in genetic studies, which estimate divergence times based on genetic differences, is criticized for its assumptions about mutation rates and the impact of genetic drift, potentially leading to inaccurate timelines.

» **Interpretation of Genetic Data:** The interpretation of genetic data in the context of human evolution also faces scrutiny. While genetic studies have revolutionized our understanding of human origins, critics caution against over-interpreting this data. There are concerns about the extent to which genetic similarities and differences can be translated into clear narratives of human migration, interaction, and speciation. The complexity of gene-environment interactions and the role of epigenetic factors further complicate the picture, suggesting that genetic data alone may not provide a complete understanding of human evolutionary history.

The critiques of mainstream theories of human evolution call for a reexamination of how we construct narratives about our past.

Hominid Fossils

The journey to understand human evolution has been significantly shaped by the discovery and analysis of fossils. These fossils provide critical insights into our evolutionary past, revealing the physical attributes, behaviors, and possible environmental adaptations of our ancestors. Key fossils have become landmarks in the study of human evolution theory.

Lucy (Australopithecus afarensis)

One of the most famous hominid fossils is "Lucy," an Australopithecus afarensis discovered in 1974 in Ethiopia. Dating back approximately 3.2 million years, Lucy's remains are significant for their evidence of bipedalism, a key trait distinguishing early hominids from their ape ancestors. Her relatively small brain size, combined with the adaptation for walking upright, provides crucial insights into the evolutionary path that led to modern humans.

Homo neanderthalensis (Neanderthals)

Neanderthals, a distinct species or subspecies of archaic humans, lived in Europe and parts of Western Asia from about 400,000 to 40,000 years ago. Neanderthal fossils have been instrumental in understanding the diversity within the human lineage. These fossils show a robust physique, adapted to the cold climates of the Pleistocene, and evidence of a sophisticated culture, including tool use and possibly symbolic behavior.

Homo erectus

Homo erectus fossils, found in Africa and across Asia, date from about 1.9 million to 140,000 years ago. This species is significant for its long tenure on Earth and its spread over a vast geographical range. Homo erectus is noted for its larger brain size compared to earlier hominids and is often associated with significant technological and behavioral advancements, such as the use of fire and more complex tools.

Other Significant Discoveries

Other notable hominid fossils include Homo habilis, often considered one of the earliest members of the genus Homo, and the more recently discovered Homo floresiensis and Homo naledi. Each of these discoveries has added depth to our understanding of human evolution, providing evidence of the diversity and adaptability of hominid species.

These fossils, while invaluable, are also the subject of intense debate and scrutiny. The interpretation of their features, and what they signify about the capabilities, behaviors, and evolutionary status of these hominids, is often challenging. Gaps in the fossil record and the difficulty in precisely dating these finds add further complexities to reconstructing the human evolutionary tree.

Critiques on the Interpretation of Hominid Fossils

While hominid fossils are invaluable to our understanding of human evolution, the interpretation of what these fossils tell us is subject to significant critiques and debates. These critiques highlight the limitations and complexities involved in deciphering our evolutionary history from fossil remains.

Limitations in Fossil Interpretation

One major critique lies in the interpretation of fossil morphology. Determining behavior, cognition, and even exact placement within the human lineage based on bone structure can be speculative. For instance, while Lucy's (Australopithecus afarensis) skeletal structure suggests bipedalism, interpreting her cognitive abilities or social behaviors from her remains is far less straightforward. This extrapolation from physical to behavioral traits often relies on assumptions that may not fully capture the complexity of early hominids.

Gaps in the Fossil Record

The fossil record is notably incomplete. This sparsity of remains leads to gaps that make it difficult to construct a continuous narrative of human evolution. Significant gaps often lead to debates about the direct ancestry of Homo sapiens. For example, while Homo erectus is often considered a direct ancestor of modern humans, the exact lineage is not definitively established due to gaps in the fossil record.

Debate over Evolutionary Significance

The evolutionary significance attributed to certain fossils is another point of contention. The discovery of Homo floresiensis, for example, challenges the conventional view of human evolution as a linear progression of increasing brain size and complexity. The small brain size of Homo floresiensis, coexisting with modern humans, suggests a more complex evolutionary scenario than previously thought.

Controversy Over Dating Methods

The dating of hominid fossils is critical to placing them in the evolutionary timeline. However, the methods used for dating, such as radiometric dating or stratigraphy, can sometimes yield conflicting results. Discrepancies in dating can lead to debates about the age of fossils and, consequently, their role in the human evolutionary timeline.

Interbreeding and Species Distinction

Recent genetic studies have revealed interbreeding between Homo sapiens and other hominids like Neanderthals and

Denisovans. This interbreeding challenges the traditional view of distinct hominid species evolving separately. It raises questions about the criteria used to define species in the human lineage and whether the concept of distinct species accurately reflects the complexity of human evolution.

In summary, while hominid fossils are crucial to theories of human evolution, their interpretation is not without challenges. Critiques focus on the speculative nature of extrapolating behavior from morphology, gaps in the fossil record, debates over evolutionary significance, controversies in dating methods, and the complexities introduced by interbreeding.

Lucy and Homo Erectus Critiques

There are critiques from those who do not believe that Lucy (Australopithecus afarensis) and Homo erectus are related to modern humans, and who question the use of these fossils as evidence for human evolution. These critiques typically focus on the interpretation of the fossil evidence and the methodologies used in paleoanthropology.

Here are some of the arguments and critiques commonly presented:

Lucy (Australopithecus afarensis) Critiques

Morphological Interpretation: Critics argue that the physical features of Lucy could be interpreted as resembling a non-human primate more than a human ancestor. They point to characteristics like her small brain size, arm-to-leg ratio, and features of the pelvis and feet that, according to these critics, suggest she was more adapted to life in the trees than to bipedal, human-like walking.

Fossil Reconstruction and Completeness: There are claims that the reconstructions of Lucy's skeleton are speculative, particularly given that the fossil remains are incomplete. Critics question the accuracy of these reconstructions in depicting her posture, locomotion, and overall appearance.

Gaps in the Fossil Record: Some argue that the gaps in the fossil record make it difficult to draw a direct evolutionary line from species like Australopithecus afarensis to modern humans. They contend that without a complete and continuous fossil record, such connections are more conjectural than evidential.

Homo erectus Critiques

Species Variability and Distinction: Critics of the significance of Homo erectus in human evolution often highlight the variability within the fossils classified as Homo erectus. They argue that this variability might indicate a collection of different species or subspecies, rather than a single line of human ancestors.

Technological and Behavioral Evidence: While Homo erectus is associated with certain technological advancements, critics argue that the presence of tools and evidence of fire use does not necessarily correlate with direct ancestry to modern humans. They suggest that these could be independent developments of different hominid species.

Interpretation of Anatomical Features: Skeptics also dispute the interpretation of Homo erectus features as precursors to modern human traits. They argue that features such as cranial capacity and facial structure might not directly indicate a lineage leading to Homo sapiens.

GENETIC ANALYSES –

MITOCHONDRIAL DNA (MTDNA)

AND Y-CHROMOSOME

In the quest to understand human evolution, genetic studies have emerged as a vital tool, offering insights into our ancestral lineage and migration patterns. Two primary types of genetic analyses – mitochondrial DNA (mtDNA) and Y-chromosome studies – have been instrumental in this quest.

Mitochondrial DNA (mtDNA) Studies: A Deeper Dive into Maternal Lineage and Evolutionary Insights

Mitochondrial DNA (mtDNA) studies have emerged as a critical tool in unraveling the complexities of human evolution, particularly in tracing the maternal lineage. These studies involve analyzing the DNA found in mitochondria, cellular structures that generate energy and are inherited exclusively from the mother. This unique mode of inheritance makes mtDNA an invaluable asset in studying maternal ancestry and evolutionary patterns over generations.

How mtDNA Studies Work

Mitochondrial DNA is distinct from the nuclear DNA found in the cell's nucleus. It contains a small, circular genome that, due to its separate evolutionary path, mutates at a different rate than nuclear DNA. Scientists analyze these mutations, or genetic markers, in mtDNA to trace lineage and investigate evolutionary relationships. The relatively high mutation rate of mtDNA makes it particularly useful for studying recent evolutionary events and human migrations.

One of the pivotal applications of mtDNA studies in human evolution has been the attempt of tracing a common maternal ancestor, often referred to as "Mitochondrial Eve." This figure, thought to have lived in Africa around 200,000 years ago, represents a point from which all modern humans can theoretically trace their maternal lineage. The concept of Mitochondrial Eve has been instrumental in supporting the Out-of-Africa theory, suggesting a recent African origin for all modern humans.

Contributions and Limitations

Mitochondrial DNA studies have shed light on various aspects of human evolution. They have provided insights into the timing of key migrations and the relationships between different human populations. For example, mtDNA analyses have helped map the spread of humans out of Africa and into other continents, illustrating the genetic diversity and migratory patterns of early human populations.

However, these studies are not without their limitations and have been subject to critical scrutiny. One major challenge is the assumption of a consistent mutation rate, which is vital for accurately estimating divergence times. Critics note that fluctuations in mutation rates can lead to discrepancies in dating evolutionary events.

Furthermore, while mtDNA provides valuable insights into maternal ancestry, it represents only a small part of the overall genetic picture. Relying solely on mtDNA to construct narratives of human evolution can lead to an incomplete understanding, as it excludes the paternal lineage and the vast majority of the genetic information contained in nuclear DNA.

Critical Perspective

Critics also caution against over-extrapolating from mtDNA data. While mtDNA can indicate lineage and ancestry, it does not necessarily provide a comprehensive view of historical populations' lifestyles, cultures, and interactions. The focus on a single maternal line also oversimplifies the intricate web of human ancestry, potentially overshadowing the broader genetic contributions from the multitude of ancestors in an individual's lineage.

In summary, mitochondrial DNA studies offer significant insights into the maternal aspects of human evolutionary history, but they must be interpreted with caution and understood as part of a larger genetic and historical context. A comprehensive approach to human evolution requires integrating mtDNA findings with nuclear DNA studies and other scientific evidence to paint a more complete picture of our origins and development.

Y-Chromosome Studies: Unraveling Paternal Ancestry

Y-chromosome studies have become a pivotal tool in the field of genetic anthropology, aimed at deciphering the paternal lineage of human populations. The Y-chromosome, unique to males, is passed relatively unchanged from father to son, making it a valuable marker for tracing paternal ancestry and understanding evolutionary changes over generations.

Mechanism and Application of Y-Chromosome Studies

The basis of Y-chromosome studies lies in analyzing genetic variations known as markers on the Y-chromosome. These markers, which include short tandem repeats (STRs) and single nucleotide polymorphisms (SNPs), mutate at predictable rates. By examining these mutations and their frequencies in different populations, scientists can infer lineage relationships and migration patterns.

One of the primary applications of Y-chromosome studies is to trace the paternal lineage and to map the migratory routes of ancient human populations. Scientists use the Y-chromosome to reconstruct paternal lineages and to estimate the timing of divergence between lineages. This approach has provided insights into early human migrations out of Africa and the subsequent spread of populations across the globe.

Challenges and Skeptical Views on Y-Chromosome Studies in Human Evolution

While Y-chromosome studies have provided valuable insights into paternal lineages and human migration patterns, several challenges and critiques have emerged, questioning the reliability and interpretations of these studies.

A foundational assumption in Y-chromosome studies is the constancy of mutation rates over time. These rates are crucial for estimating the timing of divergences and migrations in human history. However, skeptics point out that mutation rates can vary significantly due to environmental factors, genetic contexts, and random fluctuations. This variability can lead to substantial inaccuracies in timeline estimations. Critics argue that relying on these rates to draw definitive conclusions about human evolutionary events may lead to oversimplified or even misleading narratives.

Y-chromosome studies, by their nature, offer a singular perspective focused exclusively on the paternal lineage. Critics argue that this narrow focus can provide a skewed view of human evolution. It overlooks the contributions of the maternal lineage and the broader genetic interplay that shapes populations. The emphasis on paternal ancestry might ignore critical aspects of human history and evolution that can only be understood by considering the entire genetic makeup, including mitochondrial DNA, autosomal DNA, and the complex interactions between these genetic elements.

There is also a critique regarding the overinterpretation of Y-chromosome data in constructing narratives of human evolution and migration. Skeptics caution against the tendency to draw direct correlations between genetic patterns and specific historical or prehistorical events. They argue that such interpretations often involve speculative leaps, filling gaps in the archaeological and fossil record with genetic data that may not provide a complete picture. This concern is particularly pronounced in scenarios where genetic findings are used to support elaborate theories of human migration and settlement that lack corroborating evidence from other scientific disciplines.

Cultural and Environmental Overlook

Another critical viewpoint highlights how Y-chromosome studies might overlook the cultural and environmental factors that play a significant role in human evolution. Human history is shaped not only by genetics but also by cultural innovations, environmental changes, and complex socio-ecological interactions. By focusing primarily on genetic data, Y-chromosome studies may undervalue the impact of these non-genetic factors in shaping human populations and their evolutionary trajectories.

The discovery of interbreeding between Homo sapiens and other hominids, like Neanderthals and Denisovans, has further complicated the interpretations of Y-chromosome studies. These findings challenge traditional views of distinct, separate lineages and suggest a more intertwined human ancestry. Critics of the Y-chromosome methodology argue that it may not adequately account for these complexities, raising questions about the nature of species boundaries and the evolutionary processes that led to modern humans.

In summary, while Y-chromosome studies have significantly contributed to our understanding of human evolution, they face a range of challenges and critiques. These critiques emphasize the need for caution in interpretation, advocating for a more comprehensive approach that integrates genetic findings with archaeological, anthropological, and environmental evidence to construct a more nuanced understanding of human evolution.

Challenges in Genetic Studies

Despite the insights offered by genetic studies, there are inherent challenges and limitations. One major issue is the assumption of a constant mutation rate, critical for estimating divergence times in both mtDNA and Y-chromosome studies. Critics highlight that mutation rates can vary significantly, which may lead to inaccurate estimates of when key evolutionary events occurred.

Another point of skepticism is the extrapolation of genetic data to broader narratives of human evolution. While genetic studies can reveal patterns of ancestry and migration, they cannot fully illuminate the cultural, environmental, and behavioral aspects of human history. Furthermore, the focus on genetic data sometimes overshadows the importance of archaeological and fossil evidence, leading to a potentially skewed understanding of our past.

Recent revelations about interbreeding between Homo sapiens and other hominins, like Neanderthals and Denisovans, have added complexity to the human evolutionary story. While such findings challenge earlier models of human evolution, they also raise questions about the nature of species distinctions in our evolutionary past. Skeptics of strict evolutionary models argue that these interbreeding events indicate a more intertwined human ancestry than previously thought.

REASSESSING OUR UNDERSTANDING: KEY TAKEAWAYS AND THE PATH FORWARD

Summary of Main Themes

In "Rethinking Science - Questioning Evolution and the Theories," we embarked on a critical journey through the complex world of evolutionary biology, genetics, and human evolution. The book meticulously dissected and scrutinized these fields, shedding light on the intricate processes that underlie the diversity of life and the origin of species. Central to our exploration were the major themes of evolutionary theory, the intricate dance of genetics, and the nuances of human evolution, each viewed through a lens of constructive skepticism.

We delved into the traditional narratives of evolutionary theory, often portrayed as a seamless and straightforward progression of life from simplicity to complexity. However, by adopting a skeptical viewpoint, we uncovered layers of complexity and ambiguity within these narratives. The book challenged the conventional wisdom on evolutionary mechanisms like natural selection, gene mutation, and the gradual accumulation of genetic changes, inviting readers to question the sufficiency and probability of these processes in explaining the vast tapestry of life.

In discussing genetics, we confronted the enigma of genetic complexity and the controversies surrounding gene duplication, chromosomal changes, and the increase in genetic information. The book presented a nuanced critique of these genetic phenomena, highlighting the challenges and unanswered questions that they pose to the evolutionary paradigm.

The discourse on human evolution was particularly compelling, as we explored the evolutionary trajectory of our own species. This exploration was marked by an examination of the fossil record, genetic evidence, and anthropological findings, all of which were weighed against the backdrop of skepticism, revealing a story far more intricate and less certain than traditionally presented.

The Role of Skepticism in Science

This book underscored the pivotal role of skepticism in scientific inquiry. Skepticism, often misconstrued as mere doubt or cynicism, is in fact a fundamental driving force in the pursuit of knowledge. It compels us to question, to probe deeper, and to not accept explanations at face value. By questioning established theories, we pave the way for scientific progress, uncovering new insights and perspectives.

The contributions of traditional evolutionary theories to our understanding of the natural world are undeniable. These theories have provided a framework that has guided centuries of scientific exploration and discovery. However, as this book has demonstrated, these theories are not without their limitations and blind spots. It is only through continual reassessment, critical analysis, and open-minded exploration that we can hope to refine our understanding and uncover new truths.

In this spirit, "Rethinking Science" does not seek to diminish the achievements of evolutionary biology but to enrich and expand them. It invites readers to embrace the complexity and uncertainty inherent in scientific exploration, recognizing that our understanding of the world is ever-evolving, just like the species and processes we study.

Challenges and Controversies Highlighted

In "Rethinking Science," we have navigated through a terrain rich with challenges and controversies that underpin the theories of evolution, genetics, and human evolution. This exploration has not only illuminated the complexities of these scientific fields but also brought to the forefront a series of critical debates and skeptical arguments.

Evolutionary Theory Under the Microscope

The book delved deeply into the foundational principles of evolutionary theory, only to uncover layers of complexities and contentious issues. One significant challenge highlighted is the concept of gradual complexity. The traditional view of evolution posits a slow, incremental build-up of complex traits and structures. However, this book raised critical questions about the practicalities and probabilities of such a process. The improbability of a series of advantageous mutations occurring in just the right sequence to build complex structures, such as the eye or the wing, was a point of intense scrutiny.

Furthermore, the rarity of beneficial mutations posed another significant challenge. Mutations are often neutral or detrimental, and the likelihood of a mutation leading to a substantial evolutionary advantage is, by nature, quite low. This rarity casts doubt on the efficacy of mutations as the primary driving force behind the vast diversity and complexity of life.

Genetic Complexity: A Realm of Questions

The discussion of genetic complexity in the book revealed a landscape filled with more questions than answers. Gene duplication, a mechanism touted for creating new genetic material, was examined skeptically, particularly regarding its ability to lead to functional, beneficial changes. The complexities involved in integrating new genetic information into an existing, highly regulated genomic framework were emphasized, highlighting the delicate balance required for such integration to contribute positively to an organism's survival and reproductive success.

Human Evolution: Bridging Gaps and Questioning Narratives

Human evolution, a topic of immense interest and debate, was not spared from critical analysis. The book revisited the fossil record, a primary source of evidence for human evolutionary history, pointing out significant gaps and missing links. These gaps in the fossil record pose challenges to the smooth narrative of human evolution, raising questions about the completeness and accuracy of our current understanding. The book encouraged readers to consider these gaps not as mere anomalies but as potential windows into a more complex and nuanced evolutionary history.

Concluding Thoughts

As we reach the culmination of our journey through "Rethinking Science," we pause to reflect on the intricate and often contested landscape of evolutionary theories that we have navigated. This expedition has not only been an exploration of scientific concepts and evidence but also a profound exercise in critical analysis and open debate. It has been a journey that underscores the dynamic and continually evolving nature of scientific understanding.

The Value of Critical Analysis and Open Debate

Throughout this book, we have engaged with evolutionary theories not as static truths, but as hypotheses and frameworks ripe for questioning and scrutiny. This approach is emblematic of the scientific method itself — a method that thrives on challenge, skepticism, and rigorous testing. By questioning and critically analyzing the established narratives of evolution, genetics, and human origins, we have ventured beyond mere acceptance, delving into the realms of deeper understanding and intellectual growth.

This process of inquiry and debate is the lifeblood of scientific progress. It propels us to uncover new data, revisit old assumptions, and, sometimes, rewrite the narratives that we have long taken for granted. The discussions and skeptical viewpoints presented in this book are not endpoints but starting points —

invitations to continue the exploration with a curious and questioning mind.

Encouragement for Continued Skeptical Engagement

As readers close the final pages of this book, the journey does not end. The exploration and questioning of evolutionary theories, like all scientific endeavors, is an ongoing process. The complexities and mysteries of life's evolution, the intricate dance of genetics, and the fascinating story of human development are far from being fully unraveled.

I encourage readers to carry forward the spirit of skepticism and inquiry. Engage with scientific topics, not just with acceptance, but with a healthy dose of critical thinking and open-mindedness. Question, explore, and debate — these are the actions that drive the pursuit of knowledge.

Let this book be a catalyst for that way of thinking, inspiring a continued engagement with the wonders and questions of our natural world. May your journey in the pursuit of scientific understanding be ever curious, ever questioning, and ever enriching.

OVERVIEW OF THE PROBLEMS AND CRITIQUES OF EVOLUTION THEORIES AS DISCUSSED

These points highlight the range of criticisms and debates surrounding evolutionary theory, emphasizing the need for ongoing scrutiny, open-mindedness, and critical examination in the scientific community.

1. Microevolutionary Mechanisms and Macroevolutionary Patterns: The current models of evolution may not fully capture the nuances and complexities of life's evolutionary journey, suggesting the need for more comprehensive theories.

2. Interpretation of New Fossil Discoveries and Genetic Analysis: Debates revolve around the extent to which genetic evidence can inform about early human migration patterns and population structures, with concerns about over-reliance on genetic data.

3. Critique of Mainstream Theories of Human Evolution: Mainstream theories are critiqued for oversimplifying

human evolution and for overemphasizing certain aspects of the fossil record, such as cranial capacity, at the expense of other significant aspects.

4. Limitations of Contemporary Scientific Knowledge: Early criticisms of evolutionary theory stemmed from the limitations of contemporary scientific knowledge and empirical evidence.

5. Role of Genetic Constraints and Developmental Biology: Critiques focus on the oversimplified view of the genome in the Modern Synthesis and the limitations of developmental processes in producing viable phenotypes.

6. Contemporary Criticisms and Challenges: The application of evolutionary theory in fields like medicine and psychology has been criticized, with arguments that it may overlook crucial factors and fail to adequately explain complex human behaviors and social structures.

7. Suddenness of New Forms in the Fossil Record: Debates about the tempo of evolutionary change, with alternative theories like Goldschmidt's "hopeful monsters" and the concept of punctuated equilibrium by Gould and Eldredge challenging the gradual change proposed by Darwin.

8. Broad Application of Evolutionary Theory: Criticisms arise from the theory's application to non-biological systems, diluting its scientific meaning and leading to its perception as a universal explanation for a wide array of phenomena.

9. Methodological Concerns in Studying Human Evolution: Critiques include the reliance on morphological analysis of fossil remains and the use of molecular clock techniques in genetic studies, which might lead to speculative conclusions and inaccurate timelines.

10. Questioning the Explanatory Power of Evolutionary Theory: Notable scientists have raised concerns about the limitations of Darwinian explanations, especially in explaining phenomena like "punctuated equilibria" and the role of random mutation.

11. Challenges to the Modern Synthesis: Criticisms include its inability to fully explain macroevolutionary changes and the limitations in accommodating new findings in genetics, development, and ecology.

12. Historical Critiques During the Formative Years of Evolutionary Theory: Figures like Louis Agassiz and St. George Mivart raised objections based on empirical evidence and questioned the mechanism of natural selection in explaining complex structures.

13. Debates on the Role of Natural Selection and Survival of the Fittest: Critics argue that natural selection cannot account for all aspects of evolution and that the concept of "survival of the fittest" is tautological.

14. Gradualism and Irreducible Complexity: The gradualist perspective of natural selection is challenged for failing to account for the observed irreducible complexity in biological systems, and the role of random genetic mutations is debated.

15. Critiques of Epigenetic Contributions to Evolution: Debates exist about the extent to which epigenetics can influence evolution and the stability of epigenetic modifications across generations.

16. Limitations in the Explanatory Scope of Natural Selection: Criticisms include the persistence of biological features and behaviors that do not offer any survival advantage, challenging the notion that natural selection accounts for all forms of biological diversity.

I am sure you will find this book interesting and informative.

If so, please consider giving it a review on Amazon and perhaps

mention it or link to it on your social media

Thanks, it will be highly appreciated

NOTABLE KEY FIGURES

Gregor Mendel

Often known as the father of genetics, Mendel's work on inheritance patterns in pea plants laid the foundation for the field of genetics. His discoveries were pivotal to understanding how traits are passed down through generations, which later became integrated into evolutionary theory as part of the Modern Synthesis.

Charles Darwin

Darwin's formulation of the theory of natural selection in his book "On the Origin of Species" is a cornerstone of modern evolutionary biology. His observations and conclusions about the nature of species variation and the process of natural selection are fundamental to the field.

Alfred Russel Wallace

Wallace is co-credited with Charles Darwin for the independent conception of the theory of natural selection. His extensive fieldwork in the Amazon River basin and the Malay Archipelago provided crucial insights into the geographic distribution of species.

Stephen Jay Gould

Gould, known for his theory of punctuated equilibrium, argued that the fossil record does not present a smooth and gradual transition of one species evolving into another. He also raised concerns about the limitations of Darwinian explanations for phenomena like "punctuated equilibria".

Stephen Jay Gould, along with Niles Eldredge, proposed the theory of punctuated equilibrium, challenging the traditional Darwinian concept of gradual change in the fossil record. They argued that species remain stable for long periods, punctuated by rapid evolutionary changes.

Lynn Margulis

Margulis criticized the neo-Darwinian emphasis on random mutation and natural selection, proposing symbiosis as a significant evolutionary process.

Lynn Margulis was known for her theory of endosymbiosis, which emphasizes the role of symbiotic relationships in the evolution of eukaryotic cells. She critiqued the traditional Darwinian focus on random mutation and natural selection.

Michael Behe

Behe, an advocate of Intelligent Design, argued against the Darwinian mechanism of evolution through his concept of "irreducible complexity".

Michael Behe is a biochemist and Intelligent Design proponent known for his concept of "irreducible complexity." He argues that certain complex biological structures cannot be explained by Darwinian evolution.

Karl Popper

Popper initially criticized Darwinian evolution for its lack of falsifiability but later revised his stance.

Karl Popper, a philosopher of science, initially described Darwinian evolution as a "metaphysical research program" and not a testable scientific theory. However, he later revised his views, recognizing evolution as a scientifically testable theory, though he maintained some reservations about its status.

Louis Agassiz

Agassiz opposed Darwin's theory, arguing that the fossil record did not support gradual changes.

Louis Agassiz, a renowned paleontologist and geologist, was a critic of Darwin's theory of evolution, specifically arguing against the gradual change in species as suggested by Darwin, based on his interpretation of the fossil record.

St. George Mivart

Mivart raised objections to Darwin's theory, including questioning how natural selection could favor the initial stages of complex structures like the eye.

St. George Mivart was an English biologist who initially supported Darwin's ideas but later critiqued them. He particularly questioned the ability of natural selection to explain the development of complex structures like the eye.

William Paley

Paley's teleological argument was often referenced in opposition to evolutionary theory.

William Paley, predating Darwin, argued for the divine design of living organisms, a view often contrasted with evolutionary explanations post-Darwin.

Lord Kelvin (William Thomson)

Kelvin criticized evolutionary theory based on thermodynamics and the Earth's age, which he believed was too short for evolutionary processes to occur.

Lord Kelvin argued against the long timescales required by Darwinian evolution, estimating the Earth's age to be much younger than what would be needed for evolution as proposed by Darwin.

Adam Sedgwick

Sedgwick, a mentor to Darwin, did not accept Darwin's theory, arguing for divine creation based on the fossil record.

Adam Sedgwick, a geologist and priest, was one of Darwin's mentors. Despite their friendship, Sedgwick did not accept the theory of evolution, favoring a view of divine creation.

Theodosius Dobzhansky, Ernst Mayr, Julian Huxley

They were architects of the Modern Synthesis, integrating Darwinian evolution with Mendelian genetics.

These figures were central to the development of the Modern Synthesis, which combined principles of natural selection with Mendelian genetics to provide a more comprehensive framework for evolutionary biology

Thomas Kuhn

A philosopher of science, Kuhn is known for his book "The Structure of Scientific Revolutions." He argued that scientific progress is not a linear accumulation of knowledge but proceeds through "paradigm shifts." While not directly opposed to evolution, his ideas have influenced debates on how scientific theories, including evolution, change over time.

Francisco J. Ayala

An evolutionary biologist and philosopher, Ayala has critically discussed the application of evolutionary theory in areas such as ethics and religion. His work often explores the limits of what evolutionary theory can explain about human nature and society.

Sir Fred Hoyle

An English astronomer famous for opposing the Big Bang theory, Hoyle advocated for a steady-state model of the universe. His stance contributed to significant debates in cosmology, a field closely related to theories about the universe's origins.

Paul Feyerabend

A philosopher of science, Feyerabend was known for his critical views on scientific methodology and the philosophy of science. His principle of "epistemological anarchism," which argues that there are no methodological rules that are always used by scientists, has implications for debates on evolutionary theory and other scientific paradigms.

David Berlinski

A critic of evolutionary theory, Berlinski is a senior fellow at the Discovery Institute's Center for Science and Culture, a hub of the intelligent design movement. He has written extensively critiquing Darwinian evolution and advocating for intelligent design.

John C. Sanford

A geneticist and advocate of intelligent design, Sanford has criticized certain aspects of evolutionary theory, particularly the concept of genetic entropy, which argues against the long-term viability of biological systems under evolutionary processes.

Philip Johnson

A law professor and a leading figure in the intelligent design movement, Johnson has criticized the methodological naturalism of modern science and has argued for the serious consideration of intelligent design as an alternative to evolutionary explanations.

Alfred Wegener

Known for proposing the theory of continental drift, Wegener faced significant skepticism and criticism from the scientific community of his time. While not directly related to evolution, his experience reflects the challenges faced by theories that go against established scientific thinking.

MORE BOOKS FROM THIS AUTHOR

Boost Your Child's Self Esteem and Confidence: Raising Confident Kids: Your Ultimate Guide

Improve Your Child's Potential with Boost Your Child's Self Esteem and Confidence

In the ever-changing world of today, self-esteem and confidence are cornerstones upon which our children's future happiness and success are built. This is not just a book; but a powerful guide designed to empower parents with knowledge and tools needed to nurture and elevate these vital attributes.

Why It Matter

Understanding the profound impact of self-esteem and self-confidence on a child's development is the first step toward equipping them for a fulfilling life. Research has consistently shown that children with healthy self-esteem and confidence are more likely to excel academically, build meaningful relationships, and embrace future challenges with resilience. They're better equipped to navigate the complexities of today's world, make sound decisions and lead happy lives.

A Comprehensive Guide to Empowerment

The book dives deep into the psychology behind self-esteem and confidence, offering you profound understanding of these concepts. Parents gain insights into the significance of self-esteem in shaping their child's emotional well-being, relationships, academic achievements, and aspirations.

Practical Strategies

This book goes beyond theory; it provides actionable strategies and research-based techniques parents can readily apply. Whether you're a parent seeking practical solutions or a teenager eager to harness transformative principles, this guide offers a blueprint for nurturing self-esteem and confidence effectively. Each chapter is packed with advice, examples, and hands-on exercises.

What You Will Discover:

- » Delve into the psychology of self-esteem and self-confidence, differentiating between these two vital traits
- » Learn how to create an environment that fosters growth, resilience, and a positive mindset
- » Discover effective parenting practices, such as praising effort over outcomes and setting realistic expectations
- » Empower your child to explore their interests, passions, and values, paving the way for self-identity
- » Equip your child with valuable skills to handle challenges, manage emotions, and build resilience
- » Foster healthy relationships and communication skills that last a lifetime

- » Address the impact of media and social pressures on body image and self-acceptance
- » Teach your child how to set achievable goals, celebrate successes, and maintain a
- » Cultivate positive self-talk and self-affirmations. Boost self-belief

Empower Your Child for a Bright Future

Yes! "Boost Your Child's Self Esteem and Confidence" leads you through the complex web of self-esteem and self-confidence, providing practical tools to lay a solid foundation for your child's future. By the end of this journey, you'll have the knowledge and tools to empower your child to become a resilient, self-assured individual ready to conquer challenges, seize opportunities, and thrive in every aspect of life.

Invest in your child's future today! Give them the gift of self-esteem and self-confidence with this invaluable guide. Discover the transformative power of understanding, nurturing, and boosting these essential attributes, and set your child on a path to a life filled with self-assuredness, happiness, and endless possibilities.

Order "Boost Your Child's Self Esteem and Confidence" now and embark on a journey of growth and empowerment that will shape your child's destiny.

Are you a parent? GET THIS BOOK

"Useless, but Interesting Facts: A Collection of Curious Curiosities"

Discover a World of astonishing trivia

Prepare to embark on an extraordinary journey through the labyrinth of human knowledge, where the bizarre meets the bewildering and the strange becomes strangely captivating. In "Useless BUT Interesting Facts", you'll delve into a treasure trove of mind-bending information that's sure to leave you both amazed and really entertained.

Unearth the secrets behind history's quirkiest anecdotes, explore the depths of the world's oddities, and unravel the mysteries of our peculiar planet. This book is your passport to an alternate reality of the utterly useless YET incredibly fascinating.

Inside, you'll find:

The surprising stories behind everyday objects

Bizarre behaviours of humans

Weird and wonderful world records

Cultural curiosities from around the globe

.... And so much more to astonish and amuse!

Prepare to be captivated by the unexpected, dazzled by the unusual, and enchanted by the enigmatic. "Useless BUT Interesting Facts" is your ticket to a world where the ordinary transforms into the extraordinary, where the trivial becomes a treasure, and where knowledge is anything but useless.

Get ready to journey into the astonishing realm of the intriguingly